THE ARMY IN VICTORIAN SOCIETY

STUDIES IN SOCIAL HISTORY

Editor: HAROLD PERKIN

Professor of Social History, University of Lancaster

Assistant Editor: ERIC J. EVANS

Lecturer in History, University of Lancaster

For a list of books in the series see back endpaper

THE ARMY
IN VICTORIAN
SOCIETY

Gwyn Harries-Jenkins

Department of Adult Education
University of Hull

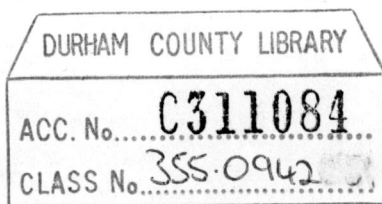

LONDON: Routledge & Kegan Paul

TORONTO AND BUFFALO: University of Toronto Press

First published in 1977
in Great Britain
by Routledge & Kegan Paul Ltd
and in Canada and the United States of America by
University of Toronto Press
Toronto and Buffalo
Printed in Great Britain by
Redwood Burn Ltd
Trowbridge and Esher
Copyright © Gwyn Harries-Jenkins 1977

RKP ISBN 0 7100 8447 1
UTP ISBN 0-8020-2263-4

To Ina, Sîona, Morag and Elaine

Contents

Acknowledgments

A book of this nature owes a considerable debt to a number of historians and sociologists who unwittingly have contributed to this study of an army during a particular period. My particular gratitude is due to my colleagues Morris Janowitz, Jacques van Doorn, Correlli Barnett and John Jackson for their advice, criticism and invaluable help. I also owe a special debt to Andrew Wheatcroft and my series editor Harold Perkin who undertook the onerous task of reading and commenting on the whole of the manuscript.

I wish to express my appreciation of the courtesy and helpfulness of the staffs of the Central Library, Ministry of Defence, the London Library, the University of East Anglia Library and the Brynmor Jones Library of the University of Hull. I would also like to thank Mrs Wendy Bray and Miss Lesley Stock for their typing and secretarial assistance.

Finally, I would like to thank my wife, not only for her advice and criticism but also for her patience and support during the preparation of this book.

Gwyn Harries-Jenkins

1

The Impact of Defeat

In the space of a few months between December 1899 and
February 1900, the last chapter in the history of the
Victorian army was written. To a shocked and incredulous
British public, accustomed to reading about the colonial
triumphs of their heroes, Roberts, Wolseley and Kitchener,
the defeats of 'Black Week' and Vaal Krantz were disasters
as great as any experienced by the British Army. There
was little to applaud in the news from South Africa. The
main British army of 47,000 men under the command of
General Sir Redvers Buller, a red-faced Devon squire, had
stumbled from crisis to crisis. In December 1899 Lord
Methuen, sent with a strengthened division to relieve
Kimberley, had been repulsed at Magersfontein. General
Gatacre, who had been sent with a brigade to clear the
Boers from the north of Cape Colony, had been defeated
at Stormberg. The commander himself, trying to relieve
Ladysmith, had been so badly beaten at Colenso on the
Tugela River that he had ordered the beleaguered garrison
to surrender.
    To a very large extent the British public over-reacted
to these tactical reverses. By European standards, the
number of casualties suffered by the army in South Africa
were small. Even at Spion Kop in January 1900 when Buller
was again badly mauled no more than 1,700 men were lost.
What magnified the scale of these disasters was not, however, the number of casualties, but the realization that
they had been inflicted by a part-time army of Boer farmers. From the beginning, the public had been led to believe that the war would be another colonial triumph.
Before the outbreak of hostilities, both press and politicians had looked forward to another easy campaign.
Writing to his mother on the eve of mobilization, George
Wyndham, the Under-Secretary of State for War and a former Guards officer, assured her that the army was more

efficient than at any time since Waterloo.(1)   In 'Blackwood's Magazine', a writer concluded his appraisal of the
coming war with the assured statement that 'no greater
mistake can be made then to suppose that the conquest of
the Transvaal Boers, left to themselves, is a task which
would severely test the British army, or which would involve an expenditure which need in the least degree alarm
the taxpayer'.(2)   Nor had the military élite been any
less optimistic.   Buller, in a final interview with Lord
Lansdowne before he left to take command of the British
troops in Natal, had confidently concluded that he would
begin his advance about two days before Christmas, and
that it would probably take him 'one month to pass
through the Orange Free State, and after that fourteen
days to get to Pretoria'.(3)   Mafeking shattered this
feeling of smug complacency.   The defeats of Magersfontein, Stormberg and Colenso ended a wave of jingoistic
optimism.   They brought home to the British public, with
dramatic effect, the realization that the Victorian military system had been found terribly wanting.   For most of
the century, from the Battle of Waterloo onwards, the
system, as an important instrument of imperial expansion,
had removed the burden of world power from the British
people as a whole.   It had allowed them the opportunity
to promote a 'Pax Britannica'.   It had pioneered the
growth of British institutions overseas, and had ensured
stable conditions for the expansion of trade.   Now in
1900, the reputation of that imperial army had been
shattered by a small force of 'slinking' Boers who, unlike the Sudanese, 'did not stand up to a fair fight'.(4)
A 'mob of good marksmen'(5) had somehow put an end to a
complete way of military life.   It was all too evident
that the Victorian army, as a fighting force, would be irrelevant in any future major campaign against the well-
equipped mass army of a European power.
   What, then, was this Victorian military establishment?
What was the relationship between it and the parent
society within which it functioned?   Why had it persisted for so long in an era of striking technological developments?   Few immediate answers to these questions
emerged from the inevitable enquiry which, by tradition,
followed every major British military reverse.   The Royal
Commissioners, under the chairmanship of the Earl of Elgin,
analysed the shortcomings of the British performance in
detail.   Their Report, published in four large volumes,
acknowledged that the whole military system as it stood
in 1899 had been tested by the war in South Africa.(6)
The Report's comments were pungent and scathing.   It
revealed causes of failure which were remarkably similar

to those published after the Crimean débâcle — the want
of organization, the lack of professionally trained officers, the inferior qualities of soldiers who were the
sweepings of cities, the mindless rigidity of a rank and
file whose tactical training had been based on the rules
of the eighteenth-century drill book. It indicted, in
its inquiry into the administrative defects and their
causes as revealed by the war, almost every aspect of the
army and the military system. Yet the Report, as comprehensive as it was, could only touch briefly on the
more fundamental causes of this military weakness. Like
a regimental history or campaign study, the Elgin Commission was primarily concerned with the analysis of a
single event or series of events within a specific time
scale, and it did not pretend to provide answers to the
wider questions which arose.

Some indication of these causes of weakness could be
seen in a comparison of the Victorian army with the mass
armies of the European powers. On the Continent, armed
forces were already becoming a world in themselves,
characterized by a separate profession, closed organization, their own value-system and norms, a special technology and their own system of law. In short, these were
the armies of an industrialized society, armies which
apparently were able to exploit the advantages of mobility,
fire power and concentration which a technical age had
conferred upon them. In contrast, the Queen's army was a
heterogeneous collection of regiments and corps, each of
which sought to maintain its own identity. They were
linked not by any universally accepted code of military
values, but by the civilian interests of their members.
A common acceptance of the standards and norms of the
English ruling class from which the bulk of the officers
were recruited, formed the basis of their attitudes towards such critical factors as the development of professionalism, the effects of technological innovation and
the growth of the civilian bureaucracy. Socially the
army, like early Victorian society, was held together by
the bonds of deference. In common with the lower orders
who had habitually deferred to their 'betters' in an
earlier and more rural society, soldiers accorded officers the respect due to rank and title. The military
code of obedience was supplemented by a complex pattern
of social relationships which mirrored those of the
parent society in an earlier period. In turn, junior
officers normally deferred to their more seniors, not
because of the latter's professional expertise but because the considerable self-confidence and authoritarian
style of general officers reflected their upper-class

assumption of an inborn right and duty to lead others.
This claim to hereditary authority was seldom questioned within the enclosed and exclusive world of the
officers' mess. Where it was challenged was in the increasingly competitive world of the expanding middle
classes. This was a group who were aware of their potential power to a greater degree than in the past. Their
discontent with some aspects of aristocratic society, in
combination with their concern for working-class poverty,
heightened their middle-class self-consciousness. Consistently throughout the century, they opposed aristocratic idleness and privilege with a fervour which reflected their concern with the puritan values of hard work and
dedicated commitment.(7) Now, at the end of the Victorian
period, it appeared as though the army were the last
bastion of neo-feudalism. Time and time again, these
critics discovered ample evidence of the way in which the
army was apparently an outmoded relic of an earlier
period that had disappeared in the wider society. The
courage, recklessness and physical toughness of many
officers, for example, reflected qualities which in the
Regency period had been cultivated for their own sake
and which in the early Victorian period had been relished
as the eccentricities of the foxhunting squire. But as
the nineteenth century ended, qualities such as these
appeared to compare very unfavourably as the hallmarks of
military efficiency, with the cool detached professionalism of the Prussian officer corps.
The nature of the middle-class attitudes, attitudes
which were a significant commentary on the relationship
between the Victorian army and society, were moulded by
the opportunity given to the public at large to serve in
the armed forces. In the absence of conscription or the
mass army, few members of society had any experience of
military service in the regular forces. Even after 1870
when the employment of troops in an expanding empire led
Cardwell, the Secretary of State for War, to introduce an
ambitious programme of military reform, the regular army
establishment was only 135,000 officers and men. In
contrast the Prussian field army alone in the 1870 war
comprised 462,300 infantry and 56,000 cavalry, while the
total effective strength of the military was well over a
million men. Shortly before the war in South Africa, the
numerical strength of the British Army compared very unfavourably with that of her Continental neighbours. In
the German Empire, with a total population of 61,479,901,
the peace strength of the army, exclusive of troops employed in the African protectorate, was a cadre of
591,507 officers and men. In France, a population of

80 million supported an army of 573,743, whereas the
British Empire with a total population of nearly 400
million produced a regular army of no more than 248,076
troops. Although this military participation ratio was
slightly improved in Britain by the opportunity given to
civilians to serve with the militia and other auxiliary
forces, it was nevertheless axiomatic that only a very
small percentage of the total male population had any
experience of army life. Some of the effects of this were
inevitable. The resulting cultural and physical remoteness of the Victorian army was an important characteristic
of its relationship with the parent society. To many
Victorians who lacked first-hand knowledge of military
life, their army was an institution whose values differed
from those of the population at large. Since it was an
organization which few of them joined, they mistrusted its
apparently privileged position during a century of change.
In particular, they were very critical of the life-style
of the officers where the daily routine within a regiment
seemed to reflect the social life of the country's upper
classes at an earlier period of history. The early
socialization of these officers, their education at a
small number of select boarding schools and their subsequent military training — or the lack of it — apparently
created a privileged group who seemed to be out of touch
with, and out of sympathy with, the technological and
social changes which had affected the remainder of
society. In contrast with the Prussian officer corps
which was aristocratic but professional, British officers,
with a few notable exceptions, seemed to be aristocratic
and amateur.

To the British public as a whole, the army was thus an
unknown institution. Most of the soldiers were in any
event serving overseas, from the Shetlands in latitude 60
degrees north to the Falklands in latitude 55 degrees
south. Their presence at home was not particularly
noticeable. And when they were seen in Britain on parades
or on guard-mounting ceremonies or on manoeuvres, they
were objects of distant admiration. Their presence was
welcome, provided it was at a distance. Closer contact
was less acceptable. There were occasions, perhaps all
too frequent, when civilian society totally rejected the
military — as when a soldier was prevented from riding in
an omnibus, or when three sergeants were expelled from a
box in Her Majesty's Theatre in the Haymarket because they
were wearing the uniform of the British Army.(8)

This ignorance of army life did not inhibit criticism
nor did it necessarily invalidate the attempts made by
the public at large to understand why their army had

failed in South Africa. It did, however, produce a number
of paradoxical situations. Consistently the public,
though willing to share in the victories of its army, were
strangely indifferent to the condition of the military.
'Over and over again when attempting to improve Army
conditions we have had to speak to a practically empty
House, empty Press gallery, and unsympathetic public.'(9)
More than two centuries of national hostility to any organized military force, a hostility engendered by Cromwell's dictatorial role, had produced a public mistrust of
a standing army in peacetime. This hostility was, as
Fortescue stresses, truly 'national', shared by all sorts
and conditions of men. It vented itself in the neglect
and maltreatment of the army by the government of the day,
and in the hatred and scorn of the people at large.(10)
For most of the nineteenth century, the middle class refused to join the officer corps, whilst the industrial
and rural worker looked upon a son who joined the army as
a disgrace to the family name; yet both groups resented
the privileges of the military, and each, in its own way,
deplored the apparent exclusiveness of army life. When
its army was defeated, the public reacted violently.
When the army was successful, the public basked in the reflected glory, pouring adulation — and financial rewards -
on its victorious generals; a paradox which led a foreign
observer of the British scene to comment, 'How this blind
glorification and worship of the Army continues to coexist
with the contemptuous dislike felt towards the members of
it must remain a problem in the national psychology'.(11)
     Equally paradoxical was the effect of British conser-
vatism. Whilst it was very evident that the army had been
found wanting, the wish to cling fiercely to old institutions meant that the public at large strenuously set its
face against whatever seemed to endanger sanctified traditions. This conservatism was still more loth to part
with a tradition if it were a famous one. The British
Army under Wellington had gained victory upon victory in
the Peninsular War. It had routed Napoleon at Waterloo.
Under leaders such as Campbell and Havelock it had sub-
jugated an immense Empire. It had waged more or less
successful war in every corner of the globe. Why then,
change in time of peace the finest army in the world?
What might be suitable for a Continental army was not
necessarily suitable for the British Army which, it was
argued, was in a class of its own. The army had stood by
itself in the past and would continue to do so in the
future. Professionalism and a dedicated commitment to
military life, like universal conscription, smacked too
readily of a militaristic spirit which was totally foreign

to the British way of life. Yet it was the lack of professionalism and the absence of this dedication to the
military career which in time of defeat were two of the
most pungent criticisms levied against the Victorian army.
British conservatism, pride in past successes and a
sense of natural independence were some of the subjective
reasons which justified the paradoxical retention of an
outmoded military system in an era of technological and
political change. But these reasons could be supplemented
by the practical realization that the geographical insularity of Britain favoured the development and growth of
the Royal Navy. To the supporters of the 'blue water
school' and to other critics of the army, it appeared as
though there was nothing carried out by the military in
the United Kingdom, which could not be more efficiently —
and more cheaply — provided by auxiliary and reserve
forces. In this context, the creation of county and
borough police forces, although it relieved the army of
much irksome duty, also removed one of the basic reasons
for the retention of a standing army in peacetime. If the
duty of preserving law and order could be transferred from
the army to a quasi-military force, then it could also be
argued that the task of providing defenders in the event
of invasion by a foreign power could similarly be transferred to a force of civilians in uniform. Yet underlying
these immediate subjective and objective reasons for the
evolution of the Victorian army in a particular form, was
the more fundamental question of the place of the army in
nineteenth-century England. In criticisms which were made
of the military, too much attention was paid to details of
superficial and peripheral importance, and too little to
the realities of the relationship between the army and
society. Conclusions which were reached were often based
on inadequate evidence, and in many ways it looked as
though the relationship of the military establishment to
society had to be based on one of two polar extremes. On
the one hand, it seemed as though the army and society
were two completely separate institutions. Britain, it
appeared, was a non-militarist nation. There was no indication of the preponderance of the military in the state.
Instead, 'civilianism' was the order of the day, a characteristic which implied a rejection of military values,
militancy and the adulation of the military ethos. This
further encouraged the public in their belief that, in the
absence of any extensive control by the military over
social life, both military objectives and organization
were wholly or partly subordinated to a civilian way of
life. A lack of public interest and involvement in military or para-military activities confirmed the distinc-

tion between 'militarism' with its apparent addiction to
drill and ceremonial or its worship of useless trappings,
and 'civilianism' with its preference for dynamic commercial and industrial development. Military activities were
thus believed to be on the periphery of societal development, and their characteristic features were seen to be
indicators of group attitudes and behaviour, rather than
the reflection of the ethos of society as a whole.(12)

    This conclusion encouraged a section of the British
public in their belief that the military was a functional
body which existed quite separately from all other institutions in society. Their attitudes were confirmed in
contemporary studies of the military. The majority of
these were critical in their approach, emphasizing the
neo-feudal characteristics of the military organization,
but this particular analysis of the military was not
limited to the works of reformers such as Trevelyan or
De Fonblanque.(13) Studies which related the position of
the military to that of other institutions in society
tended to concentrate on circumstances and occasions when,
as in the Crimea, there were differences of opinion between the political and military élites. Much of what was
written examined differences as to the control and direction of war operations with a view to restating the in-
stitutional arrangements which were in existence. When
more general aspects of military life were considered, as
when 'The Times' and 'Punch' in the 1850s deplored at
length instances of 'military jocularity', the generally-
reached conclusion was that these examples of hazing and
bullying were symptomatic of an institution which was different from the remainder of society. The identification
of the military, in emotive language, with such attributes as authoritarianism, inhumanity, coercion and persistent social conservatism made it difficult to reach
any other conclusion. Even the mass of campaign literature which appeared towards the end of the nineteenth
century endorsed the feeling, through glorifying selected
aspects of military life, that the army was different from
the remainder of society.(14) Nor did the writings of
Victorian officers contradict this conclusion, for the
majority of these army apologists sought to convince the
public that there was something almost mystical about life
within the military organization and that the participants
were not like ordinary men. The features of military life
were apparently completely different from those of society
at large and for the 'man on the Clapham omnibus' this
portrayal was particularly prevalent in the popular culture and mass media accounts of Victorian service life.
The common characteristic of these diverse publications

was thus the emphasis that was placed on those qualities
of the Victorian military which suggested that the army
could be distinguished from comparable civilian structures. Indeed, descriptions of the army tended to per-
ceive this as a total institution in which clearly defined barriers had been erected between the organization
and the remainder of society. This perception then encouraged the belief that interaction between members of
the military establishment and the world at large was
extremely limited, a conclusion which drew attention
again to the differences which apparently existed between
the military and the parent society.

Since most contemporary Victorian literature on the
subject of the army sought to draw attention to the way
in which it differed from other institutions in society,
it is easy to forget that this was only one point of
view. In contrast with these writings, the work of other
theorists emphasized the similarities between the army and
the remainder of society. In some ways Gaetano Mosca's
classic work 'Elemanti di Scienza Politica', which appeared in 1895, brought out most clearly the extent to
which Victorian soldiers were men of their time. Here,
officers as part of a ruling class were seen to be closely linked to other members of that class through ties of
kinship, shared educational experiences, a common life-
style and a mutual wish to ensure the preservation of the
status quo. 'Army officers', declared Mosca, 'retain
close ties with the minority which by birth, culture and
wealth stand at the top of the social pyramid.'(15) This
point of view thus stressed that the army was an integral
part of society. It emphasized that armed forces were
subject to the same internal and external pressures as
any other large-scale organization within society. Military social forces and army professionalism were closely
related to those of other occupational groups, for the
army in common with these groups, was forced to adapt to
increasingly important technological developments within
society.(16) The members of the military could not be
differentiated from other members of the wider social
structure. The Victorian officer was a representative of
that broader society. His ideological attitudes and political commitments were the same as those of his civilian
counterpart. Like them, he was a Conservative or a
Liberal, or less frequently, a Radical; not the member
of an exclusively military political party. He belonged
to the same political clubs — the Carlton or the Reform —
or in common with other members of the ruling class was
a member of prestigious social clubs. There were insti-
tutional parallels in the social organization of military ·

and civilian structures. Although in the twentieth
century much of the military social organization appears
to be out-moded and a relic of the Victorian period in
which it was formalized, no such criticism can be generally made of the army in the nineteenth century. Some aspects of this structure did indeed tend to persist in the
military after they had disappeared in civil society. At
the beginning of the Victorian period in particular, it
was evident that the characteristics of a Regency England
with its emphasis on 'Honour' had lingered on in the military establishment. This was evidenced, for example, by
the eagerness with which officers were prepared to take
part in a duel to wipe out an alleged insult long after
duelling had disappeared from the scene in civil society.
In 1840 the Earl of Cardigan was tried before the House of
Lords for fighting a duel with Captain Harvey Tuckett formerly of the Earl's regiment, the 11th Hussars. Three
years later Lieutenant-Colonel David Fawcett of the 55th
Foot was killed in a duel by his brother-in-law, Lieutenant Alexander Munro of the Royal Horse Guards.(17) But
increasingly as the century progressed, this legacy of
an earlier age disappeared, until the military became an
organization and a profession whose members were integrated individuals within the total societal context. In
short, when looking at the Victorian army, we are from
this point of view also looking at the broader picture of
Victorian society.

Ultimately, attitudes towards the relationship of the
Victorian military and the parent society fluctuated between these two extremes. To many Victorians, their army
was an institution whose values differed from those of the
larger society. It was an organization which few of them
joined. They reacted unfavourably to its apparent privileged position during a century of social and technological change. The physical and cultural remoteness of the
military encouraged a commonly held belief that the army
was, in some way, an aberration within British society. It
was, at best, a necessary evil whose existence in peacetime could only be justified in terms of its imperial
functions. These were functions which were superfluous
within the immediate territorial vicinity of the parent
society, and as the century progressed any apparent justification for the retention of the army within the United
Kingdom as a means of preserving law and order lost its
validity.

Conversely, other Victorians saw the armed forces as a
reflection of dominant social values. It was an institution whose existence ensured the preservation and continuation of basic norms and standards. The latent

function of the army was to provide credibility for the
State as an independent and autonomous society. It was
the existence of this army which propped up the structure
of society, for the role of the military was essentially
the support of an established social order. And, in
carrying out this task, the Victorian army mirrored in its
attitudes, its rituals and its way of life, the culture of
an upper-class élite who dominated that society.

The question of the effect of the relationship between
the army and society on the effectiveness of the Victorian military was not one which could be easily an-
swered. Evidence could be produced to justify a number
of points of view. Many of the reached conclusions were
value judgments, their expression illustrating the moral
or political commitment of the debater. Diverse view-
points frequently generated considerable discussion as the
harshest critics of the Victorian army variously claimed
that the military was either too isolated from, or too
enmeshed with, civil society. The significant point was
that the general feelings which were developed in discussions of the relationship of the military and society
reflected deeply held internalized values. These, themselves, inhibited rational and objective assessment, but
there were other factors which also affected the evaluation of the relationship between the army and Victorian
society, and any analysis of this relationship has to take
these into account. Some of these are particularly important. They not only explain further why the performance
of the Victorian army was so poor, but they also reflect
accurately the dominant social values of Victorian
society. Ultimately it is the nature of these values and
the way in which changes in the parent society were not
necessarily paralleled in the army, which does much to
explain the position of the military in nineteenth-century
England. Yet these values did not originate in a vacuum,
and to look more critically at the Victorian army it is
also necessary to consider further the values and attitudes of society as a whole.

2

Officer Recruitment

One of the areas which brings out very clearly the complexities of the relationship between society and the
Victorian army is that of officer recruitment. This was a
problem area which affected very considerably the position
of the army in three dimensions — as a profession, as an
organization and as a political force. Professionally,
the question which arose was whether a system of recruitment which allegedly produced a self-perpetuating clique
would coincidentally hinder the development of professionalism, or whether the latter would only be ensured if
recruitment were 'open', that is, based on criteria which
emphasized the need for individual merit and ability. As
an organization, the army was faced with the problem of
developing a viable relationship not only between the members of the officer corps and their subordinates, but also
between the military and political élites. In both instances, the crucial question was whether the association
between 'officer' and 'gentleman' was so vital to the development of this effective relationship that it precluded
the widening of the base of recruitment to include candidates from the 'lower orders'. In terms of its political
attitudes, the Victorian army was similarly faced with
problems which were derived from the adoption of any
policy governing officer recruitment. On the one hand, it
could be argued that a policy based on achievement produced a politically sterile military force, an argument
which has been succinctly summed up by a modern commentator:(1)

> It implies that there is nothing in the professional
> soldier's social background which would endanger in-
> ternal democracy. If the officer corps were a repre-
> sentative cross-section, they would hardly harbour
> intentions to upset the political balance. They could
> not be accused of imperial ambitions beyond those

sanctioned by the popularly elected legislators.
On the other hand, an argument similar to that advanced
by Mosca in 'The Ruling Class' suggested that the recruitment of the officer corps from a limited area which
coincidentally provided recruits for the political élite
ensured that the interpenetration of these élites reduced
the probability of inter-group conflict.

The military was not alone in facing these problems.
Similar difficulties arose, for example, when attempts
were made to lay down policies governing recruitment into
the Home Civil Service or the Indian Civil Service, and to
a considerable extent the situation in the army was
paralleled by that prevailing in Victorian society as a
whole. This was particularly noticeable in the way in
which solutions which were put forward both for the military and for other institutions in the parent society were
affected by attitudes towards the opposing concepts of
open competition and patronage. Supporters of either
system could advance convincing arguments in favour of the
adoption or retention of their point of view. Open competition, it was argued, ensured by its policy of recruiting candidates on the basis of their academic ability that
corruption, lack of incentive, and ascriptive bias were
swept away.(2) Professionally, it would create an occupation based on achievement. Structurally, it would develop a meritocracy. Politically, it would limit the
power of an entrenched ruling class. These were important
considerations in the context of the changes which were
taking place in Victorian society as a concomitant of technological advancement, and the arguments which were advanced by 'Reformers' such as Trevelyan and Macaulay were
very persuasive. But there was another side to the picture. Patronage, it was argued, had brought into the public service at an early age men of ability such as Pitt.
Their nomination to high office had furthered their
careers and had enabled them to assume command when young.
How otherwise, it was contended, would Wellington as a
forty-year-old Lieutenant-General in 1808, last on the
list of promotions, have been appointed over the heads of
his seniors to command in the Peninsula? Patronage, moreover, had created group loyalty and encouraged a feeling
of cohesiveness. Politically, therefore, it had ensured
the recruitment and promotion of those who had an interest
in the preservation of the system. Structurally, a system
of patronage had discouraged deviancy. Professionally, it
had brought into organizations the amateur whose attitudes
and actions were not motivated by considerations of career
or personal advancement.

In many ways the difference between these points of
view was a difference of emphasis. In neither case did
the supporters of one system or the other adopt an extreme stance. Open competition was not equated with open
recruitment. Barriers were still raised against candidates
from the major part of society. Patronage was controlled
both by convention and by more formal rules, so that the
system was not entirely open to large-scale abuse. When
there was evidence of deviation from these rules, political and public reaction was most marked. The outstanding example in the nineteenth century was the revelation
that from 1804 to 1806 there had existed more or less
openly a traffic in commissions in which Mary Ann Clarke
the mistress of the Duke of York, then Commander-in-Chief,
had played the major role. For over two years, she used
influence with the Duke to obtain army commissions and
promotions for her clients whom she charged fees of between 200 and 1100 guineas. In 1809, six months after
Mrs Clarke had separated from the Duke, Colonel Wardle,
MP for Okehampton, moved in the House of Commons that an
inquiry be made into the behaviour of the Commander-in-
Chief with respect to promotions, the disposal of commissions, and the raising of new levies, and, after a
brief debate, the Government agreed that the inquiry should
take place. The inquiry terminated on the 23 February
after several witnesses, including Mrs Clarke, had given
evidence before a Committee of the House of Commons. While
a subsequent six-day debate in the House exonerated the
Duke, public opinion led the Duke to tender his resignation of the chief command of the army, a resignation
which George III was pleased to accept. The difference
of emphasis which characterized the more fundamental arguments about open competition and patronage was that while
supporters of the former paid particular attention to the
impact of the system on the development of professionalism,
supporters of patronage tended to emphasize the political
advantages associated with their choice of system.
    This difference of emphasis was brought out clearly in
the conflicting attitudes which were adopted toward the
composition of the officer corps in the Victorian army.
For supporters of a system of patronage, it was assumed
that a gentleman, however imprecisely or crudely defined,
had a definite social role to play within society. As
Burn points out,(3)

    The country had to have its gentlemen to make and . . .
    administer its laws, to officer its armed forces, to
    conduct its diplomacy, to fill the episcopal bench
    and to do a score of other things. Although a parti-
    cular member of the class might be sluggish, timid,

corrupt or illiterate, gentlemen as a whole were
credited with enough public spirit, probity, courage
and education to make them the essential servants of
the Crown.

Once this premise was accepted, then from it followed
the belief that the individual officer would carry out
his duties in his own way, according to the believed
standards of an English gentleman, without worrying about
external evaluation of his military professionalism. This
seemed to be confirmed in a number of instances. Kinglake
summed up a popular feeling when in commenting upon Lord
Raglan's conduct at the Battle of Alma in making a personal reconnaissance into the Russian lines, he wrote,
'The horseman who rode his hunter across the valley of
the Alma and momentarily give it its head, was not an
ideal personage but a man of flesh and blood, with many
very English failings.'(4)   In these and similar comments,
it is apparent that for many commentators, it was far pre-
ferable for the officer to be a gentleman, notwithstanding
his many failings, than for him to be a cad whose professionalism would be matched by his mercenary attitudes.
Thus Wolseley, one of the few truly 'professional' generals to emerge in the Victorian period, was said by Lady
Geraldine Somerset to be 'No gentleman', whereas her kinsman, Lord Raglan, was described as 'an honourable man and
gallant to a fault'.(5)   Civilian assessment of officers'
performances continually tended to emphasize the merit of
subjective concepts of 'honour', 'bravery' and 'temperament', all of which drew attention to qualities of character which were believed to be those of a gentleman, rather
than stress the need for officers to possess positive
military skills.   The gesture of General Gough on the
second day at Ferozeskah during the Sikh War of 1845, in
riding out in a white coat to draw the enemy's fire away
from his soldiers, epitomized the concept of gentlemanly
chivalry.(6)   It illustrated the courage and temperament
which, with his racy Irish brogue, endeared Gough to his
men, but it also raised questions about the wisdom of a
commander-in-chief exposing himself to such risk.   Yet
Gough's courage was seen to be beyond approach.   It was
the continuation of an attitude which had led Marlborough
to ride slowly up and down the line at Blenheim to hearten
his men to be passive under fire.   It was a foretaste of
Raglan's behaviour at Alma.   But it was rarely asked
whether these qualities of character were those which were
demanded from officers, particularly senior officers, in
the complex military of a technological age.

At times, this reluctance to criticize the professional
capabilities of army officers was so marked, that to the

modern-day student of the Victorian military establishment
it appears as though expected behavioural attitudes were
derived from social group membership to the complete exclusion of occupational considerations. Certainly this is
a conclusion which can be drawn from a statement made by
the Adjutant-General, Sir John MacDonald, in 1840:(7)

> It is the proud characteristic of the British Army that
> its officers are gentlemen by education, manners and
> habits; that some are men of the first families in
> the country, and some of large property, but the rules
> and regulations of the service require strictly that
> they should conduct themselves as ought gentlemen in
> every situation in which they may be placed.

To a large section of Victorian society, this preference for the recruitment as officers of gentlemen who were
unconcerned by external evaluation of their professionalism could be readily rationalized. The existence of a
narrowly-based socio-economic pattern of recruitment into
the army could be justified on the grounds that the
average Englishman would not accept as an officer anyone
who was not his social superior, a thesis which specifically precluded considering the professional ability of
the military officer. This was clearly brought out in a
letter from Sidney Herbert to Mr Raikes Currie on 25
September 1857:(8)

> In despotic countries, the strong military feeling
> induces military obedience . . . here . . . military
> obedience would be impossible, were it not that the
> soldier comes from the class that is accustomed to
> respect and obey the class from which the officer
> comes.

Sidney Herbert's subsequent comment in the House of
Commons in 1860 further developed this point and it clearly summarized the attitude of those people who defended a
system of patronage on the grounds that the army needed
gentlemen in the officer corps:(9)

> I am one of those old fashioned persons who believe
> that gentlemen officers are a great advantage. In
> this country, which is not a military nation, I am not
> ready to give up anything which tends to secure to our
> officers a ready and willing obedience.

This argument that discipline appeared to depend for
its effectiveness on the believed superiority of the ruling
class, also seemed to be confirmed by the preferences of
the ordinary soldier. 'I know from experience', wrote
Rifleman Harris, 'that in our army the men like best to be
officered by gentlemen, men whose education has rendered
them more kind in manners than your coarse officer, sprung
from obscure origins, and whose style is brutal and over-

bearing.'(10)   This was an attractive argument since it
glossed over the self-interest which led these supporters
of patronage to seek the preservation of the system. Additionally, its persuasiveness could be reinforced not only
by reference to the inescapable evidence of the way in
which this army, officered by gentlemen, had expanded an
Empire throughout four continents, but also by recalling
the warning given by the Duke of Wellington in 1828:

> The description of gentlemen of whom the officers of
> the Army were composed made, from their education,
> manners and habits the best officers in the world, and
> to compose the officers of a lower class would cause
> the Army to deteriorate.

Irrespective of the validity of this assumption that
patronage brought into the army men of distinction who
were committed to the preservation of certain normative
standards, this attitude toward the pattern of recruitment was important for a number of reasons. It suggested,
inter alia, that the military preferred to emphasize the
importance of neo-feudal concepts such as bravery and
honour rather than ideas of ability and merit which were
more applicable to an organization in a period of considerable technological development. It implied that
organizational norms — obedience, discipline, certainty —
were more important than professional elements which emphasized the significance of training, education and
devotion to study. In addition it stressed the very
important point that the military officer, like his
counterpart in other sections of the public service, was
a gentleman whose choice of a career was motivated by his
wish to serve his country for honour rather than for personal reward. All of these were suggestions which because
of their emphasis on high-flown moral virtues appeared to
negate those criticisms of the military which saw it simply
as a class based élite in which ascription rather than
achievement was the foundation of success.

In contrast, the supporters of competition argued from
the premise that a restricted pattern of entry into the
military produced a concomitant lack of professionalism.
As the 'Quarterly Review' argued in 1848:

> It is reasonable to assume that officers coming chiefly
> from the higher and middle walks of life, have re-
> ceived in their youth the ordinary education of
> gentlemen. But in what walk of civil life can people
> get into positions of importance on the mere assump-
> tion that, being respectably born, they must have been
> duly educated?

The individual who approached his occupation with attitudes
of an amateur gentleman, did not, it was argued, have any

very strong incentive to take his profession seriously. He
was, it appeared, a typical member of an officer corps
who in common with his eighteenth-century counterparts in
France and England, did not have a military career in the
way in which people in other occupations devoted their
whole life to the pursuit of a professional activity.(11)
Officers, according to their own interpretation of the
military image, performed a service which had been the
traditional function of the ruling class, 'that of the
warrior who protects civil society'. This service however did not constitute a full-time occupation. It was
part of a whole range of social activities embracing the
management of estates, participation in national and local
administration, family business and other élitist obligations which bore no direct relation to military activities.
To carry out these non-military responsibilities, officers
interrupted their service with frequent leaves, expecting
both in peace- and war-time to return at regular intervals
to civilian life. 'The real defect of the system of purchase', commented the 'Saturday Review', 'consists in its
tendency to encumber the army with amateurs and to relax
the ties which bind the officer to his profession.'(12)

It was however a criticism which was not directed at
the military alone. The Administrative Reform Association,
under the chairmanship of Samuel Morley (1809-86), equally
attacked the Civil Service, arguing in its 'Official Paper
No. 1' in May 1855 that 'the whole system of Government
Office is such as in any private business would lead to
inevitable ruin'. The conclusion that patronage encouraged the entry into the British public service of amateurs
whose attitudes toward their professional responsibilities
left much to be desired was an important comment on this
traditional method of recruitment. But in attacking systems of patronage both in the military and in the parent
society, critics were also quick to point out that there
was no guarantee that the gentleman brought to his selected occupation those altruistic qualities of character
which could be accepted as a valid alternative to professionalism. Matthew Higgins (1810-68), writing as
'Jacob Omnium', argued in 'A Letter on Administrative
Reform' that the 'Upper Ten Thousand' had 'hitherto mono-
polized every post of honour, trust and emolument under
the Crown, from the highest to the lowest. They have
taken what they wanted for themselves; they have distributed what they did not want among their relations, connections and dependents'. A similar criticism could be
made of the army where there was ample evidence of
corruption, of veniality and of an absence of high moral
standards, which suggested that the consequences of

adopting a system of patronage were too high a price for
society to pay. At the beginning of the Victorian period
in particular, the 'flâneur' in the officers' mess was not
necessarily Thomas Arnold's Christian gentleman whose
attitudes and behaviour were governed by his high moral
sense and his devotion to manly virtues. The legacy of
the Regency buck, the dandy, the man of spirit who could
break with his whip all the windows of the High Street —
fanning the daylights as he called it — was not yet dead.
In many regiments, the assured unquestioning snobbery of
the rich dominated the officers' mess. This was a world
in which a Colonel of the Guards could give Storr and
Mortimer, the Regent Street goldsmiths, £25 a quarter to
furnish him with a new set of studs every week during the
season.(13) Wealth and a title, not the charismatic
qualities of leadership subsequently ascribed to a gentleman, could give officers such as James Brudenell, 7th Earl
of Cardigan, almost unlimited power, so that after he had
been removed from the command of the 15th Hussars, he was
allowed to purchase the lieutenant-colonelcy of the 11th
Light Dragoons (later Hussars) and turn it into a socially
exclusive club.
     In this situation, critics of the military could con-
tinually refer to evidence which suggested that the commissioning of these wealthy socialites did not bring into
the regiments men of distinction. Duelling, bullying,
hazing, drunkenness, gambling and gluttony were some of
the military vices to which critics of the early Victorian
army readily drew the attention of the public. But the
military situation was particularly complex, for it
created special problems and produced specific reactions.
Despite the criticisms which could be made of these officers, it was nevertheless argued that the patronage
system at least brought into positions of military importance, men who served the State loyally because it was the
guardian of their own privileges and possessions. There
was therefore, it appeared, a continuing need to find men
who could not only, by virtue of their education and
ability, adopt a more professional attitude but who also
possessed those qualities of a gentleman which guaranteed
that they would wish to preserve the status quo. 'The
problem of army reformers', said 'The Times', 'was to
provide a body of officers who will not cease to be
gentlemen.'(14) The alternative, it was feared, was an
'imperium in imperio', a military force which, separated
by impassable barriers from the remainder of society,
transferred its allegiance from the Head of State to its
own immediate heads.(15) Yet in trying to ensure a passive apolitical military by encouraging the recruitment

of gentlemen, further difficulties arose. This assumption ensured, if it were adopted, that the military
remained a closed avenue of social mobility, an area
within which there was no place for 'The middle class ...
a class between the clergy and the legal and medical professions and the higher merchants on the one side, and the
work people . . . the great middle class who carry on all
our great industrial and marine operations'.(16)
    Yet all the arguments which were advanced either for or
against patronage and competition were based on subjective
assumptions which could not be evaluated with any degree
of objectivity. If there was no evidence to prove that
the amateur gentleman was the only person fit to be an
officer, equally there was no means of guaranteeing that
the widening of the basis of entry into the military
would bring into the establishment a successful group of
bourgeois whose ability was axiomatic. This was an area
of considerable uncertainty. Even the most avid supporters
of reform in the public service, men who were prepared to
endorse Trevelyan's contention that it was necessary to
'improve the spirit and character of the public service'
in order that 'the present period will be distinguished
above all others in this country for practical executive
improvements',(17) could not agree on the criteria of
'ability'. In particular they doubted whether evidence of
academic ability would ipso facto produce the type of individual needed in the public service. If there were
doubts about the effects of open competition in this field
of recruitment, doubts which prompted James Wilson, the
Member of Parliament for Westbury and son of a Quaker
business man, to suggest that the open competition which
Gladstone and Trevelyan wanted, 'would be productive of an
enormous amount of mischief',(18) how much greater was the
uncertainty in the area of military recruitment. Here,
the arguments put forward by reformers such as Graham, who
stressed the need for entrants to the public service to
have 'moral worth and personal merits', or the Radical Sir
Benjamin Hawes who urged the importance of 'commonsense',
or Sir George Lewis, Chancellor of the Exchequer in 1855,
who considered that 'discretion and trustworthiness' were
all-important, were of consistent importance.(19) All of
these were attributes of character which seemed to suggest
(if the integrity of the officer corps were to be preserved) the continuing need for some form of patronage
within the Victorian military establishment. Moreover,
while the force of the opposition to the existing military
system was exacerbated by the belief that it deprived the
middle class of a relatively easy means of acquiring a
coveted social standing, there was no means of finding out

whether a wider basis of recruitment would ensure that a
particular status was afforded to these new entrants. As
Thackeray in his vitriolic account of the social problems
faced by Major Ponto of the Royal Marines and his son
Cornet Wellesley Ponto of the 120th Hussars pointed out,
there were considerable difficulties which these newcomers
to the landed interest encountered.(20)   In short, this
was a complex problem which was bound up with the niceties
of Victorian social attitudes and existence of a hier-
archically-constituted society, and no ready-made solutions were forthcoming.

But was entry into the military so completely restricted to members of a particular socio-economic group? More
importantly, did a limited entry, if it existed, produce
a low level of military professionalism? Or did the
military position, even after the reforms of 1870 which
formally abolished patronage, reflect a more general
malaise in society as a whole?

The initial problem which arises in considering these
questions is that the years of the Victorian military
establishment not only covered a relatively lengthy
period in the history of the British army, but also embraced an era of considerable social and technical
change. The England of 1900 was different in a large
number of respects from the England of a hundred years
earlier. Indeed, it can be argued that there were considerable changes within the time-span of a decade. Burn,
in analysing the Victorian period half way through the
century, stressed that the England of 1852-67 was not the
same as that of 1842 or 1872. Many of the old landmarks
remained, but persistent and insidious change had altered
the landscape in considerable detail.(21)   Rostow in his
identification of the five stages of economic growth in
the life of industrial societies,(22) produces a typology
which shows how during the Victorian period the shift of
the economy, with its attendant social changes, from the
basic industries sector to consumer goods and services,
had a considerable effect on British society. Little sur-
vived unchanged. What did survive in the military establishment was not an entirely unchanging pattern of life.
The field of recruitment did not remain static. Indeed,
rigidity would have produced a military caste, completely
divorced from the remainder of society. Instead, the
army adjusted to changes in a social structure whose com-
ponents altered in character and prospered or declined in
importance, according to the pressures of industrializa-
tion. Nevertheless, a constant factor, irrespective of
changes in the social order, was the interpenetration of
the army and the landed interest, 'that great judicial

fabric, that great building up of our law and manners
which is in fact, the ancient polity of our realm'.(23)
This was a link of the greatest importance, for notwithstanding the criticisms which could be made of this landed
interest, for as long as the possession of land was the
symbol of social status and the ownership of a family seat
was the sign of financial standing, the English landed
gentry retained their dominant place in a wide variety of
activities. In national or local politics, in the Church
and in the civil bureaucracy, both in London and the
counties, this landed interest formed a ruling class which
made the major decisions for the remainder of society. In
possession of the symbols and sinews of power, this group,
despite their lack of agreement over many issues, formed
a reasonably homogeneous whole exercising both influence
and authority over the 'lower orders'. The identification
of the military with this landed interest thus integrated
the former with the wielders of power. The officer corps
was not isolated from the ruling class. Drawn from that
class, it had no need to use the ultimate source of physical power which it controlled, to advance its own corporate self-interest. No sectional interest existed to
encourage the military to think of itself as having an
identity which was different from that of the landed
interest. Their motives and attitudes were identical. The
officer thought of himself as a part of the landed interest fulfilling his obligations of public service within
the military establishment, in the same way in which a brother or other relative served in the Church or in Parliament.
     At the same time the landed interest was not a closed
group. It was a constantly changing fluid structure, the
composition of which altered as some of the old families
sank into landless oblivion and newcomers moved into the
group from outside. Equally, as the membership of the
landed interest changed to meet the pressure of new
wealth,(24) so did the pattern of recruitment into the
military alter. Relatively easy access into the landed
interest thus produced patterns of recruitment, and variations in the socio-economic background of military aspirants, which were often overlooked by critics of the
Victorian military establishment. While the connection
between the old established landed interest and the military establishment could be fairly readily traced, the
relationship of the officer corps to newcomers into this
ruling group was less easily noted. In itself, this difference was not important. What was of significance,
however, was the way in which changes in the structure of
the landed interest and, by extension, changes in recruit-

ment into the officer corps, prevented the military from
becoming an exclusive caste. It was this fluidity which
partly accounted for the difference in the attitudes of
the German Junkers and the British officer corps, since
a continuing extension of the area of recruitment from
which officers of the Victorian military were selected
hindered the development of rigid isolated attitudes.
Many of these 'newcomers', moreover, were able to make a
significant contribution to military and political life.
The 3rd Baron Gifford (1849-1911), for example, enjoyed
a distinguished military and public career. Educated at
Harrow, he joined the army in 1869 and was a lieutenant
in the 83rd Regiment in 1870. After transferring to the
24th Regiment, he served on Wolseley's staff in the
Ashanti War of 1874-5 where he won the Victoria Cross.
Following Wolseley to the Zulu War of 1879-80, he later
retired in 1882 as a major in the Middlesex Regiment (57th
Regiment). Although he had succeeded to the title in
1872, Lord Gifford did not remain long at his seat in Old
Park, Chichester, for from 1880-3 he was Colonial Secre-
tary for Western Australia and the Senior Member of the
Legislative Council, before becoming Colonial Secretary of
Gibraltar.

    On first examination, this is not atypical. This was
a family with a tradition of public service. Lady
Gifford, the daughter of General John Street, formerly of
the 57th Foot and a veteran of the 1842 China War, herself
served in the Army Nursing Service in the South African
War of 1900-2. The 2nd Lord Gifford had served in the
army for fifteen years after his succession to the title.
The family, moreover, had intermarried with the landed
interest. The 3rd Baron's mother was the Hon. Frederica
Charlotte, daughter of the 1st Lord Fitzhardinge, formerly MP for Cheltenham, who owned Berkley Castle and
20,274 acres in Gloucester with a gross annual value of
£33,717. His maternal grandmother was Charlotte, daughter
of the 4th Duke of Richmond and Gordon who from his seat
at Goodwood controlled 286,411 acres in Scotland and
Sussex.

    It was the 1st Lord Gifford who provided evidence for
the assertion that entry into the landed interest was not
impossible. The son of a grocer and linen-draper of
Exeter, Robert Gifford was created Baron Gifford of
St Leonards in Devon in 1824, on his appointment in close
succession as Lord Chief Justice of the Common Pleas and
Master of the Rolls. This example confirms the presumption that recruitment into the army was not limited to
members of the 'old landed families', and, indeed, members of these 'new' families often enjoyed a career of a

type which was rarely open to individuals in the older
landed interest whose life-style tended to be prescribed
for them. In the Gifford family, for example, the 3rd
Baron's brother, Hon. Maurice (1859-1910) had experienced
a remarkable para-military career. Educated at HMS
Worcester, Greenhithe, he started life as an officer in
the Merchant Navy but after six years he became, in 1882
during the Egyptian Campaign, 'Galloper for Mr G. Lagden,
Special Correspondent Daily Telegraph'. After this, his
wanderings took him to Canada where he was a Scout for
General Middleton in Riel's Rebellion of 1885, and to
South Africa where he was a Scout in Salisbury's column
during the 1893 Matabele Campaign. Three years later he
raised the 'Gifford Horse' during the Matabele Rebellion,
and although a serious wound led to the amputation of his
arm, this did not prevent his coming back to England in
1897 to command the Rhodesian Horse in the Queen's Jubilee
Procession. The outbreak of war in South Africa saw him
back on active service in the Kimberley Mounted Corps, a
locally raised irregular unit.

There were innumerable cases where new entrants into
the landed interest, for a variety of reasons, encouraged
members of their family to enter the armed forces. In
most instances these new men, who set out to become members of the landed interest, were very quick to appreciate
that their integration into county society depended on
their adoption of the life-style of their conservative
neighbours, and of the institutions of this society, military life was one of the easiest in which to participate.
So Albert Brassey (b. 1844) of Heythrop House, Chipping
Norton, the fourth son of the railway contractor Thomas
Brassey, followed his education at Eton and University
College, Oxford with a period of service in the 14th
Hussars, before he returned to his 4,275 acre estate. John
Hubbard, whose position in the commercial interest as a
Russian merchant and a Governor of the Bank of England,
was recognized by his elevation to the peerage as Lord
Addington during Lord Salisbury's second administration,
similarly established himself as a part of the landed interest. In 1873, before he was created a peer, he owned
2,576 acres in the Home Counties with a gross annual value
of £4,887. His son Egerton, 2nd Baron Addington (1842-
1915) followed him into the family business as a partner
in John Hubbard & Co., but his grandson and heir to the
title, Hon. J.G. Hubbard (b. 1883), served as a junior
officer in the Buckinghamshire battalion of the Oxford
Light Infantry. These examples are not unique, and,
indeed, a characteristic of the military establishment
during this period was the extent to which the army was

used as a 'confirmer' of an aspired social status. For
the man with money, land could be fairly readily bought,
in the way in which the railway speculator George Hudson
bought Londesborough Park and 12,000 acres for £470,000
from the 6th Duke of Devonshire. But the acquisition of
social status was more difficult, and for many aspirants,
a period of military service in a fashionable regiment was
one of the easier ways to secure the approval of a conservative local society. Acceptance by the military establishment often set the seal on such aspirations and gave
to the individual an entry into a closed group which, because of its intricate social network, accepted the officer into a wider social circle than that afforded through
land ownership alone.

In looking at the connection between the landed interest and the Victorian military establishment, however,
it is necessary to define more precisely the composition
of the former. The landowners headed the landed interest
and were in a position to control it, but it contained
more than them alone. As Thompson points out, the
language of classification which contrasted a moneyed or
commercial interest with the landed interest cut right
across the lines of modern class distinctions.(25) It
reflected the ethos of a society which in the earlier
period of the nineteenth century emphasized vertical divisions by sources of livelihood and occupation, rather
than horizontal divisions by Marxist concepts of class.
Thus in general terms, this interest embraced the great
body of the agricultural community, from nobleman to yeo-
man, from squire to parson, and from blacksmith to publi-
can.

But the whole edifice rested on the respect which was
given to property, and more particularly to the possession
of a landed estate. Social mobility within the inner
hierarchical structure of the landed interest was closely
associated with the acquisition or loss of the territorial
possession. Without land, there was no involvement in the
organizations and institutions of local society. Land
ownership guaranteed that the individual was bound to the
demands of a rural environment. It involved him not only
in the management of his own estate but also in more
general issues of agricultural and commercial improvements
within the community. An ordered, traditional and seemingly permanent hierarchical society was created and, in
the end, the permanence of the State and its constitution
seemed to depend on the maintenance of this landed interest. 'Has not', asked Coleridge in 1800, 'the hereditary
possession of a landed estate been proved by experience to
generate dispositions equally favourable to loyalty and

established freedom?'(26)   And this concept of stability
was not only associated with the great landowners at the
apex of the hierarchy. Bateman in analysing the holding
of the 217,049 'smaller proprietors' who owned on average
about 15 to 25 acres, hoped that the Government would
'shatter the legal bonds which fetter and prevent the free
sale of land, and the expansion thereby of Class 6
(Smaller Proprietors) — a class which, if multiplied
fourfold, would add greatly to the stability of our
fatherland and its institutions'.(27)

The relationship between the landed interest and the
military establishment was thus a welcome one, for it
seemed to ensure that the military was committed to the
maintenance of stability. It appeared that the military
officer who was coincidentally a member of the landed
interest was committed, because of this, to the preservation of the social structure which was the source and the
recognition of his differential privileges. The connection between the upper strata of the landowners and army
officers was moreover particularly welcome, for this suggested that the latter were no separate group, but were
part of an inner social pyramid in which the privileges
of ownership were balanced by the duties and responsibilities of leadership and administration.(28)

> As a general rule, the higher positions, the Lords-
> Lieutenant, Sheriffs and Justices of the Peace, the
> Grandjurors, the Commissioners of roads, the Conser-
> vators of public edifices — in fact, all that in
> France is done by the salaried servants of our variable
> Government . . . — all this, I say, is in England exe-
> cuted by the class of country gentlemen who, while they
> continue to live at home, regulate the finances and
> administer justice in their respective localities,
> spontaneously, gratuitously, and with an admirable
> degree of perfection.

At the apex of this pyramid were the landed nobility.
In general terms, it had always been held that the possession of an estate was indispensable for the support of an
hereditary title. Despite economic and social changes in
the nineteenth century, this principle held good until the
late 1880s, and indeed traces of its effect lingered on
into the next century when Haig, for example, informed the
Prime Minister that he would decline any offer of a peerage unless he received an adequate grant to enable him to
maintain a suitable position.(29)

Eventually Haig was given an earldom and £100,000, but
Bateman in 1883 was able to draw attention to some excep-
tions to the principle. In his analysis of the great
British landowners, he listed 66 peers who held small

estates, that is, less than 2,000 acres and a minimum
gross annual value of £2,000 per annum. Of these 11 owned
less than 50 acres, and a further 20 held less than 500
acres, while 60 peers from a total peerage of 585 held no
land at all. Among the owners of a small estate was the
2nd Lord Raglan of Cefntilla Court, Usk in Monmouthshire,
whose land amounted to no more than 95 acres. His father,
Field Marshal Lord Fitzroy Somerset, eighth son of the 5th
Duke of Beaufort, had been created Lord Raglan in 1852.
Initially, it was doubtful if Lord Fitzroy would accept
the offer of a peerage. The Prince Consort wrote to the
Prime Minister, 'It would be a great pity if Lord Fitzroy
were to be obliged to decline the peerage on account of
poverty. . . . Under these circumstances, rather than
leave Lord Fitzroy unrewarded . . . the Queen would herself bear the cost of the fees.'(30) One of the military
peers who was completely landless was George Drummond
(1807-1902), 14th Earl of Perth, 6th Earl of Melfort, Duc
de Melfort in the peerage of France, Hereditary Thane of
Lennox and Hereditary Steward of Menteith and Strathearn.
Drummond, who had joined the 93rd Highlanders in 1824, was
restored to his titles, attainted after 1745, by special
command and recommendation of Queen Victoria in 1853, but
after years of litigation he failed to recover the ancient
family estate of Drummond Castle which, although it had
been restored to the family in 1784, had passed through
the female line to the Earls of Ancaster. Nevertheless,
despite these exceptions, the principle generally held,
even to the extent that the Hughendon estate was purchased for Disraeli, with a loan of £25,000 from the Duke
of Portland, to give substance to his suitability as the
Conservative Prime Minister. Similarly, fortunes acquired
in industry and commerce were spent on the acquisition of
a suitable estate as the prerequisite for a peerage.
     The position of the landed nobility was supplemented by
the baronetage and the landed gentry. This group, following Bateman's adopted classification can be defined to
include the greater gentry, whose estates were from 3,000
to 10,000 acres and whose income was more than £3,000 a
year, and the squires who owned 1,000 to 3,000 acres or
whose larger estates produced less than £3,000 a year.
This group numbered almost 4,000 (1,288 greater gentry and
2,529 squires) who between them owned 12,816,970 acres
from a total acreage in England and Wales of 34,523,974
acres. When to this figure is added the land owned in
Ireland and the 5,728,979 acres owned by 585 peers, it can
be seen that some 4,000 landowners at the apex of the
landed interest owned almost half of the United Kingdom.
     The relationship between this part of the landed

interest and the military can be examined from several
points of view. Consistently during the nineteenth century, there was a marked link between the army and the
landed nobility. Of the 316 land-owning Victorian peers
who died in the decade from 1897-1916, no fewer than 139
had been members of the officer corps. This did not, however, mean that these military peers formed a necessarily
cohesive group. They included both large and small landowners. Eyre Challoner Henry Massey, 4th Baron Clarina
(1830-97), owned about 2,000 acres around his seat at Elm
Park, Clarina, Limerick. In contrast, Cromartie Suther-
land-Leveson-Gower, 4th Duke of Sutherland (1851-1913),
was the largest landowner in Britain, with a holding of
1,358,600 acres in Shropshire, Staffordshire and in
Sutherland around his principal seat at Dunrobin Castle.
Similarly the length of their military service varied
considerably. General Lord Clarina had spent the whole of
his adult life in the army from the time he was first commissioned in 1847 until he retired in 1891. The Duke of
Sutherland, on the other hand, had served with the 2nd
Life Guards only from 1870-5, although he had retained a
link with the military establishment by commanding the
volunteer regiment of the Sutherland Rifles from 1882
until he succeeded his father ten years later. But it
was not inevitable that it was only the smaller landowners who were fully committed to a military career. The
2nd Earl of Cawdor (1817-98) was a retired lieutenant-
colonel and former MP for Pembrokeshire who owned some
102,000 acres. The 10th Earl of Galloway (1835-1901)
served for fourteen years in the Royal Horse Guards until
he retired in 1869, inheriting 79,000 acres four years
later. Nor did rank in the peerage necessarily affect an
individual's involvement with the army. The 15th Marquess
of Winchester (1858-99), premier Marquess of England, who
inherited his title and 4,800 acres of land in 1887, was
a long-serving captain in the Coldstream Guards who had
been on the Nile Expedition of 1885. The 4th Marquess
Conyngham (1857-97), in contrast, retired after three
years in the army when he succeeded to the title and
167,000 acres on the death of his father in 1882.

The common link in all cases was the ownership of land,
even though the size of estates showed such considerable
variations. Apart from this, the pattern of military
careers and an involvement in politics provided a common
background which emphasized a persistent link between
membership of the landed interest, the army and Parliament both in the House of Commons and House of Lords.
Within this overall framework there were indeed innumerable variations. Some military peers came from a family

with a long tradition of army service. The 5th Earl of
Carrick (1835-1901) and the 6th Earl (1851-1909) were
cousins, both of whom served in the army. General Lord
Congleton (1839-1906) was succeeded as 5th Baron by his
son Henry (1890-1914) of the Grenadier Guards. In other
instances the relatively short period of military service
enjoyed by these peers was a part of their socialization
programme. Before the 1st Duke of Sutherland (1758-1833)
had inherited his father's marquessate and estates, his
life had followed the eighteenth-century tradition of
public school, Oxford, and the Grand Tour of Europe. His
great-grandson the 4th Duke, at the end of the nineteenth
century, preferred to include in his programme a period of
service in a fashionable regiment. In this choice he was
copied by a large number of the lesser landed interest,
some of whom used these years in the army as a means of
confirming their claim to social status and to recognition
as an established member of this privileged group.
   The variations in this relationship between the landed
interest and the military establishment can be seen in
more detail through analysing Bateman's commentary on individual estates owned by these peers and the greater
gentry. Table 1 shows the number of these landowners in
1883 who had themselves served in the military. It does
not fully explore the depth of this relationship, for it
does not take into account the kin of these landowners, so
that while Colonel Matthew St Quinton (1800-76), for
example, combined a military career in the 17th Dragoons
until his retirement in 1851 with the management of 7,000
acres, valued at £70,000 a year, in East Yorkshire, Bateman's entry simply refers to his son, William. Similarly,
the entry for the 8th Duke of Beaufort (1824-99), who had
served in the 1st Life Guards and 7th Hussars from 1841
to 1854 when he went onto half pay after succeeding to
51,085 acres worth £56,226 a year, does not mention the
other members of the family — Lieutenant-General Lord
R.E.H. Somerset, Major-General Lord Fitzroy Somerset
(subsequently Lord Raglan) and Colonel Lord J.T.H.
Somerset — each of whom gave years of service to the
Victorian Army.
   Nevertheless, while this table does not reflect the
full contribution to the military establishment which this
part of the landed interest may have made, particularly
since it was often second and third sons who joined the
army, rather than the title-holder or his heir, it is
still very pertinent. Indeed, in limiting analysis to an
examination of the relationship between a major landowner
and the army, this table emphasizes the significance of
the former's particular contribution to the officer corps.

TABLE 1 The landed interest and the army, 1883

Landowner	Regiment							
	Life Guards	Cavalry	Guards	Rifle Brigade	Line	Artillery	Engineers	Total
Peerage	43	25	47	6	16	0	0	137
Greater gentry	52	132	71	23	124	13	2	417
Total	95	157	118	29	140	13	2	554

Life Guards = 1st Life Guards, 2nd Life Guards and Royal Regiment of Horse Guards
Cavalry   = Dragoon Guards (1st–7th), Dragoons, Hussars and Lancers (1st–7th)
Guards   = Grenadier Guards, Coldstream Guards and Scots Fusilier Guards
Line   = Regiments of Foot (1st–99th)

   Seen against the background of the army as a whole,
where at any one time there were at least 6,000 officers
on full pay, this total of 554 officers is small in comparison. It suggests that the peerage and greater gentry
could only have provided a very small part of the officer
corps during any particular period, and that the majority
of officers must have been recruited from other sources.
This, indeed, appears to be confirmed by the way in which
the service of officers who are included in the composition of Table 1 covered a wide time-scale. Three of the
peers — General the Earl of Albemarle (14th Foot), the
Marquess of Donegall (7th Hussars) and the Earl of
Stradbroke (Coldstream Guards) — and six of the gentry —
Colonel William Graham of Dumfries (17th Lancers),
J. Whyte-Melville of Fife (9th Lancers), Sir Charles Munro
of Ross (45th Foot), Major R.S. Sitwell of Derby (29th
Foot), General Sir Hugh R. Ferguson-Davie of Crediton
(73rd Foot) and H.P. Delme of Fareham (88th Foot) — were
born in the eighteenth century. Their military careers
were associated with the Peninsula and Waterloo, whereas
for the younger members of this group their military service may not have started until after the Crimean War.
Thus E.A. Gore (1839-1912) of Derrymore, Co. Clare, where
he owned over 8,000 acres, joined the army in 1858. At
the time of Bateman's book he was a Lieutenant-Colonel
commanding the 6th (Inniskilling) Dragoons, though after
serving in the 1881 Boer War, Gore received further promotion until in 1900 he reached the rank of Lieutenant-
General. But while the individual experience of these
officers in 1883 could embrace campaigns from the Peninsula through to the First South African War of 1881, this
did not necessarily infringe the feeling of group homogeneity. Entry into the landed interest and into the
military was a continuing process where every year saw
the entry of new recruits. Sons succeeded their fathers
as members of both groups. Ferguson-Davie was succeeded
as the 3rd Baronet by his son, John (1830-1907), who
during his service in the Grenadier Guards from 1846-58
had served in the Crimea before he retired to live the life
of a country gentleman, combining estate management and a
love of shooting and salmon fishing with public service as
MP for Barnstaple and the Colonelcy of the Devon Militia.
Nor was he alone. In Scotland where the landed interest
had traditionally provided a succession of army officers,
Sir Robert Menzies (1817-1903) of the Black Watch was
followed as 8th Baronet by his son, Sir Neil Menzies
(1855-1910), a former Captain in the Scots Guards who continued to manage their family estate of 97,000 acres at
Castle Menzies. In England, the heir of the owner of

Stonehenge, the 4th Sir Edward Antrobus (1848-1915),
Colonel of the 3rd battalion, Grenadier Guards, who had
served on the Suakin Expedition of 1885, was his son
Edward (b. 1886), a junior officer in his father's old
regiment. This continuity of service reinforced the perpetuation of the specific ethos which gave to the officer
corps its distinctive character. It was an attitude
towards the idea of public service which bewildered many
of the critics of the Victorian army whom in stressing the
defects inherent in a system of patronage, were unable to
comprehend fully the reasons which induced members of the
landed interest to spend a lifetime of service in the
officer corps.

These military landowners had, on paper, little to
gain from their service but, continually, they left their
country estates to serve in all parts of the world, though
frequently their army career brought them very little in
the way of reward. This can be seen in many examples, and
even senior officers did not necessarily enjoy all the
plaudits or recompense which they might have expected. Of
such officers, the career of General Sir Redvers Buller
(1839-1908) was typical. Commissioned in the 60th Rifles
in 1858, Buller, 'a red-faced Devon squire', was soon
associated with Wolseley, for after serving in China in
1860 he joined the latter in the Canadian Red River Expedition of 1870 and the Ashanti War of 1874. Four years
later, after fighting in the 1878 Kaffir War, Buller won
the Victoria Cross and was promoted a brevet Lieutenant-
Colonel for his services in the Zulu War of 1878-9. At
the time of Bateman's analysis of the landed interest,
Buller, who was recorded as a member of the greater
gentry, owning 5,089 acres with a gross annual value of
£14,137 in Devon and Cornwall, appeared to be an example
of the prominence which could be achieved by a military
squire. Nor was Buller's career yet finished and his
responsibilities as a landowner did not stop him from
achieving an almost perfect Victorian success story.
After serving as an ADC to Queen Victoria, Buller's service in the Boer War of 1881 and in the Sudan from 1882-5
brought him a knighthood among the decorations which were
heaped upon him. His promotion to General and his command of British troops in South Africa during the Boer
War from 1899-1901 should have been the final heights of
a lengthy military career, but in what could have been
his moment of triumph, Buller, like Lord Raglan before
him, was abused and blamed for the defeats suffered by
the British military.

It can also be argued that only a small part of the
landed interest was actively serving in the army at any

one time, and that Table 1 exaggerates the importance of
the contribution made by this group to the composition and
ethos of the Victorian army. This is a valid criticism,
but it must be considered against the background of the
parent society. This was a period when, in the absence
of conscription in Britain, few members of society had a
direct experience of army life. In contrast, a far higher
proportion of the landed interest than of any other section or group had been members of the military establishment. No other interest or class could, at the time when
Bateman made his analysis, rival the claim that one in
four peers and one in three of the greater gentry had
served in the army. Indeed there were times and places
when the significance of this relationship was particularly marked. In the East Riding of Yorkshire, for example,
at the beginning of this period, three of the four resident peers were serving in the military: Lord Muncaster
(1802-38) of Warter Hall (10th Dragoons), Major Lord
Beaumont Hotham (1794-1930) of Dalton Park, MP for Leominster, and General Lord Howden of Spaldington Hall,
Colonel of the 43rd Regiment. The fourth peer, Captain
Lord Middleton (1769-1856) was a naval officer. Moreover,
the importance of this relationship between the landed
interest and the military establishment went a lot further
than the figures alone suggest. The group in Table 1 was
far from homogeneous. The size of estates varied from
the 201,640 acres of the Duke of Atholl (Scots Fusilier
Guards) or the 126,295 acres of Alfred Donald Mackintosh,
The Mackintosh (71st Highland Regiment of Foot) to the
3,482 Suffolk acres of Captain Sir William Parker who had
served in the 44th (The East Essex) Regiment from 1844 to
1850. Income derived from the estates similarly evidenced
considerable variations in life style and consumption
power. The Duke of Bedford (Scots Fusilier Guards) with
an annual income of £141,793 or William Legh of Lyme Park,
Stockport (21st Foot) with £45,000 a year, enjoyed an
economic power which was ostensibly superior to the
£4,500 gross annual value of the 42,000 acres owned in
Inverness by Ewen Macpherson (Cluny Macpherson) who had
served in the 42nd (The Royal Highland) Regiment from
1822-33 when he retired on half pay as a twenty-nine-year-
old captain.
    Similarly the length of military service varied consi-
derably. For some members of the landed interest, the
short time which they spent in the army, usually in a
fashionable regiment, was part of that socialization process which groomed them for their subsequent role in
society. Viscount Seaham, for example, who was the 5th
Marquess of Londonderry from 1872 to 1884, served with the

1st Life Guards from 1845-8 and then represented Durham
North in the House of Commons as a 'Liberal Conservative'
from 1847-54. An Oxford graduate, the Marquess played a
prominent part in local administration in north Wales and
northern England. A Deputy Lord-Lieutenant for the
counties of Montgomery, Merioneth and Durham, Lord Lieutenant of Durham after 1880, a magistrate in Flint and
north Yorkshire and Chairman of Merioneth Quarter Sessions, the Marquess was also a lieutenant-colonel in the
militia. In contrast, other members of the landed interest who came from families with a long military tradition,
established by several medieval crusaders and commanders,
looked upon the army as their full-time career. Of this
group, families such as the Dawnays were typical examples.
In the eighteenth century the head of the family, the 3rd
Viscount Downe, had a remarkable career. MP, FRS, Lord
of the Bedchamber to the Prince of Wales and Lieutenant-
Colonel of the 25th Foot, he died at the age of 33 in
1760 from wounds received at the Battle of Campden. The
8th Viscount, Major-General Sir Hugh Dawnay (1844-1924),
and his sons John, 9th Viscount (1872-1931) and Major Hugh
Dawnay (1875-1914) were distinguished officers who maintained a family military tradition whilst also looking
after large estates with seats at Wykeham Abbey, Dingley
near Market Harborough, Danby Lodge, Grosmont, and
Hillington Hall near King's Lynn. The family were represented in the House of Commons by two of the 8th Viscount's brothers. The elder, Colonel Lewis Payn Dawnay
(1846-1910), had served in the Coldstream Guards from
1865-79 before going into the House as Conservative Member
for Thirsk from 1880-5 and for Thirsk and Malton 1885-92.
His younger brother, Hon. G.C. Dawnay, who had fought in
the Zulu War and the Sudan, represented the North Riding
of Yorkshire from 1882-5 and was Surveyor-General of the
Ordnance from 1885-6, thus combining both a political and
military career.
    Individual differences between members of this group
were numerous, and the lack of homogeneity was an important check on the establishment of a military caste.
Nevertheless, a number of common interests encouraged the
development of a certain uniformity of attitudes. The
first common link was one generated by active service.
Seventy-eight members of this group had served in the
Crimea or the Indian Mutiny. Five of them had won the
Victoria Cross: Major-General Sir Henry Havelock Allan,
General Sir Redvers Buller, Sir William Montgomery-
Cunningham, Major Clement Walker-Heneage and Colonel Sir
Robert Loyd-Lindsay (Lord Wantage). Throughout its life,
the Victorian military establishment was almost con-

tinually engaged in some form of active service, in campaigns which ranged from Canada to New Zealand and from
China to South Africa. When Victoria came to the throne,
her army was fighting in Canada; when she died, the reve-
lations of British military weaknesses in the Second
South Africa War truly indicated that an era had ended.
In the intervening years, the Victorian army fought some
seventy-two campaigns of varying military and political
significance, and in all of these the landed interest provided officers at every level of command. Lieutenant-
General William Massy (1838-1906), for example, of
Grantstown Hall, Tipperary, whose recreations in his old
age were 'country life, gardening, building, travelling
and study' had been so badly wounded in the assault on the
Redan on 8 September 1855 that when captured by the Russians they did not take him prisoner. Twenty-five years
later he commanded the Cavalry Brigade in the Second
Afghan War, his fellow officers including the future 4th
Earl of Minto (1847-1914) who became Governor-General of
Canada in 1898 and Viceroy of India in 1905, and Charles
Euan-Smith of Manor House, Shinfield, Reading, subsequently Minister Resident at Bogota. Nor were these officers
exceptional examples. The 14th Viscount Gormanston (1837-
1907), who owned some 11,000 acres around Gormanston
Castle, Co. Meath and who was Governor of Tasmania at the
end of the century, served with the 60th Rifles during the
Indian Mutiny, while the 4th Viscount St Vincent died from
wounds in the Battle of Metamneh in 1885, his brother retiring from the Berkshire Regiment when he inherited the
title to manage their 4,500 acres at Norton Disney, Newark.
    This common link between various members of the landed
interest was also furthered by their membership of military clubs. Three hundred and ninety-one members of this
group belonged to 'Service Clubs': United Services (107),
Junior United Services (72), Army and Navy (120), Guards
(70), and Naval and Military (22). Their mutual interests,
however, were even more noticeable in their record of public service which not only emphasized a common awareness
of their responsibilities in society but also drew attention to the extent to which active or retired military
officers played their part in provincial and national administration, and in politics. In every county, serving
and retired officers assumed the responsibilities of an
English country gentleman. Nor was this limited to the
peerage alone. At Moreby Hall in the 1870s, for example,
Captain Thomas Henry Preston (1817-1906) drew £6,894 from
5,142 acres in the East Riding of Yorkshire. 'An old
Etonian, Hussar officer (7th Dragoons), J.P. in both
Ridings and Deputy Lieutenant of the East, keen agricul-

turalist and member of the Carlton Club, Captain Preston
was a well-established squire.'(31)   In Cheshire, W.J.
Legh of Lyme Park whose military career in the 21st Foot
from 1848-55 included active service in the Crimea,
carried on a family tradition of public service by representing Lancashire (South Division) in the House of Commons from 1859-65 until defeated by Gladstone, and Cheshire
(East Division) from 1868-85 as a Conservative member.
Additionally, he was a Deputy Lieutenant and magistrate of
both Lancashire and Cheshire, but his public responsibilities did not divert his attention entirely from military
activities, and from 1878-85 he commanded the yeomanry
regiment The Lancashire Hussars.

Similarly many of the squires, defined by Bateman as
the owners of the smaller estates between 1,000 and 3,000
acres, played a very important part both in military and
public life, performing valuable functions in local and
national administration, as the last link in the chain of
communication headed by the Lord Lieutenant.   There were
many examples of these military squires in nineteenth-
century rural England.   In the East Riding of Yorkshire,
for instance, the Burtons of Cherry Burton were representative of a much larger sample.   At the 1796 Beverley
election, Major-General Napier Christie Burton (1758-
1835) a veteran of the American War of Independence became Tory MP for the Borough.   In 1839 a kinsman, Fowler
Burton (1822-1904), joined the 97th (The Earl of Ulster's)
Regiment as a seventeen-year-old ensign, and in the Crimea, as a Major, was awarded the brevet of Lieutenant-
Colonel for his distinguished services in the Siege of
Sebastopol.   While his barrister brother, David, carefully doubled the return from the 1,624 acre family
estates and invested heavily in commercial enterprises,
Fowler pursued a distinguished military career until he
retired as a General in 1881.   At Hardingham Hall, Norfolk,
Major William Edwards (1855-1912), who had won the Victoria Cross at Tel-el-Kebir, combined the management of
his estate with service in the county as a justice and
Deputy-Lieutenant.   In Lanarkshire, George Colt, (1837-
1909) a retired Captain in the 23rd Foot (The Royal Welch
Fusiliers) with whom he had served during the Indian
Mutiny 'enjoyed all the usual recreations of a country
gentleman' from his estate at Gartsherrie.

Frequently, these military squires carried on a family
tradition of public service which mirrored that of their
more important neighbours in the landed interest.   Thus
Colonel Archibald Leslie, who died in 1913, 14th Laird
of Kininvie, Banffshire, grandson of General William
Stewart of Lesmurdie, combined a military career (1860-95)

which concluded with the command of the 79th Cameron
Highlanders, with service from 1868 as a JP and a Deputy-
Lieutenant, and with the management of his estates at
Kininvie and Lesmurdie. From Stoke Place, Slough, the
Howard-Vyse's provided a succession of military officers.
General Richard Vyse was MP for Beverley in 1806-7 and his
son General Richard Howard-Vyse served as the Borough's
MP from 1807-12. In the 1856 Army List the family was
represented by Lieutenant Edward (3rd Light Dragoons),
later a Lieutenant-General, Captain Francis (Royal Horse
Guards), Captain George (2nd Life Guards) and Major
Richard Howard-Vyse (Royal Horse Guards). In 1867,
Edward then a forty-one-year-old Lieutenant-Colonel in
the 3rd Hussars, married Mary Norcliffe, and in due course
their son Colonel Cecil (1872-1935) inherited the North-
cliffe's Langton Hall estate in East Yorkshire, settling
down to the life of a squire in an English rural area.
Sometimes these squires were more closely connected with
other parts of the landed interest, so consolidating a
common concern in estate management and in the affairs of
the county which bridged the difference in the size of
their land holdings and in their incomes. At Wycliffe
Hall, Barnard Castle, Captain Henry Hotham (1855-1912)
who had served in the 90th Light Infantry in the Kaffir
War of 1878, spent his life after retiring from the army
in 1893 in managing his small estate and bringing up a
family of six children. As heir presumptive to the 5th
Lord Hotham, Captain Henry was related to a family with
considerable military and landed interests, for the 3rd
Lord Hotham (1794-1870), a distinguished General who had
served with Wellington in Spain and at Waterloo, combined
the management of an expanding 20,000 acres estate at
Dalton Hall, Beverley, with a Parliamentary career as
Member for East Yorkshire from 1841-68.
     The importance of the part played by army officers in
public service at national and local level was a significant feature of civil-military relationships, and this
link between public and military service draws attention
to the way in which its importance was not simply dependent on the size of the group. What was important was
that at all rank levels, both within the military establishment and the landed interest, there was a mutual recognition of the common ethos which bound them together.
It was this ethos, reinforced by the interpenetration of
the two groups and the links of family and social networks, which was one of the reasons why officers did not,
during the lifetime of the Victorian army, feel completely
isolated from the parent society. This was a relationship
of considerable importance. It contributed in no small

measure to the retention of the pattern of traditional
authority in the military, for these officers brought
into the officers' mess the habits of a rural society.
They were accustomed to receiving personal loyalty from
their tenant-farmers, a loyalty which to a large extent
was not shaken by economic change within society. As
Thompson comments,(32)

> Even the fiercest financial bargaining between landlord
> and tenant took place within the framework of a tradi-
> tional and accepted social order. It was not that
> landlord and tenant relations were kept in separate
> compartments, economic affairs not being permitted to
> impinge on social and political matters, nor was it a
> case of landowners continually and forcefully remind-
> ing tenants of their inferiority and dependence.
> Rather it was a matter of the loyalties of a large
> family, the very paternalism to which the organs of the
> middle class so much objected. Struggles over rents,
> or games, or leases were domestic to an estate; to
> the outside world it presented a united front.

These were the attitudes of mind which were taken by
these members of the landed interest into the regiment.
The 'loyalties of a large family' were expected and received. It was the breaches of this unwritten code which
so many officers found incomprehensible, for traditional
authority demanded a display of mutual confidence and
trust. The behaviour of the Earl of Cardigan when in
command of the 15th Hussars and 11th Light Dragoons was
deplored by other members of the officer corps, not because he had been too severe or too autocratic, but because his conduct had destroyed the regiment's loyalties.
It was this concept of loyalty to the family of the regiment which was insisted upon from every new entrant into
the group, and through a process of identification and
imitation, the ethos of the landed interest was accepted
as that of the military establishment.

The importance of this link between the landed interest
and the officer corps, can be more critically evaluated
if the definition of 'landed interest' is expanded to include immediate members of the families in question.
Table 1 does not bring to light the full contribution made
to this link by extended families such as the Lindsay's,
of which Lieutenant-General J. Lindsay, grandson of the
5th Earl of Balcarres, his namesake Lieutenant-General
James Lindsay, son of the 7th Earl, and Military Secretary
to the Duke of Cambridge, Lord Lindsay (the future 9th
Earl), Colonel Robert Loyd-Lindsay (later Lord Wantage),
and his brother, Captain Sir Coutts Lindsay, were important members. This contribution can be more readily seen

when the link is examined for a single year. It can be
expected that any conclusion reached will not be constant
over a period of time, particularly during an era of
social and political change, but findings will be indicative of trends in the pattern of recruitment into the
Victorian army.

For a number of reasons, the year of 1838 can be initially selected for further examination. The date is
sufficiently distant from the end of the Napoleonic Wars
for the military to have recovered from the immediate re-
ductions which were imposed upon it when peace was signed.
Army Lists of an earlier date tend to be unrepresentative,
for they either include the names of those officers who
only joined the military to meet the requirements of the
army in a period of wartime expansion, or because restrictions on numbers recruited produced a distorted pattern of
entry. 1838 is also a year which is far enough away from
Bateman's analysis of 1883 to enable meaningful comparisons to be made. Additionally, the earlier year forms
part of a specific period of considerable social and
legislative change in Great Britain, which encourages comparison with developments in related administrative and
political areas. In this year, the British Army numbered
6,173 officers on the active list and a further 6,009 who
were on half pay. Table 2 shows the contribution made to
the former total by the 548 peers who were title holders,
and by their immediate kin, that is, brothers, sons and
nephews.(33)

TABLE 2   Interpenetration of the army and the peerage: 1838

Rank in peerage	Number	Number who held commissioned rank				
		Title holder	Son	Brother	Kin	Total
Duke	30	5	16	15	10	46
Marquess	36	4	11	11	15	41
Earl	213	18	36	85	57	196
Viscount	65	9	13	14	9	45
Baron	204	22	20	45	47	134
Total	548	58	96	170	138	462

The figure of 462 officers is approximately 9 per cent
of the officer corps, so that it can be argued that this

is a small proportion of total strength, but this must be
related to the size of the nobility. The latter, including their extended family, only comprised 13,620 out of a
total national population in 1831 of 13,896,797 and it can
be seen that their contribution was out of proportion to
the relative size of this group. It is questionable, however, whether this contribution was made generally by members of the peerage, or whether specific families exhibited evidence of an above-average military tradition.

To a certain extent, this question draws attention to
the way in which the English aristocracy was an 'open'
nobility. Despite attempts made, as in the Peerage Bill
of 1719, to limit entry into the group, the English nobility did not form a caste as did their French counterparts. In France in the eighteenth century the aristocracy had successfully excluded the bourgeoisie from the
military, the church and the civil service,(34) but no
perpetual oligarchy had been established in Great Britain.
Indeed, successful military service, in common with a
successful Civil Service career or legal career, or the
possession of political interest, was a recognized method
of entry into the peerage. The effect of this on the relationship shown in Table 2 is that the picture is
changed by the particular contribution which recently ennobled military families, or military families who had
been advanced in the peerage, made to the officer corps.
Thus Lord Paget, for example, who succeeded as Earl of
Uxbridge in 1784, was advanced in the peerage as Marquess
of Anglesey in 1815 in recognition of his services as one
of Britain's most distinguished cavalrymen in the Napoleonic Wars. The Marquess was himself a General and the
Colonel of the 7th Hussars. In addition he had three
brothers who were commissioned: Captain William RN,
General Sir Edward and Sir Charles who was a Rear Admiral
of the White. Three of Anglesey's sons were also officers: Commander Lord William Paget, Lieutenant Lord
Alfred who, on going onto half pay in 1845, became Clerk
Marshal of the Royal Household and Principal Equerry to
the Queen, and Lord George Paget (Life Guards) who subsequently commanded the Light Brigade in the Crimea.

From his principal seat at Beaudesart Park, Rugeley,
reported by Bateman to total, with other land, 29,737
acres with a gross annual value of £110,598, the Marquess
was head of a large military family. His nephew Frederick
Paget, a Lieutenant in 1838, retired as the Colonel of the
Coldstream Guards in 1851. The female Pagets also became part of a wide-spread military network. A daughter
of the Marquess married Captain George Byng (1806-86) of
the 47th Foot who subsequently succeeded as 2nd Earl of

Strafford;   a sister was married to Lieutenant-General
Sir John Erskine, and a niece was married to Colonel Hon.
Standish O'Grady of the Coldstream Guards. The family,
however, was not atypical and there are innumerable examples of the wide-spread extended family network which
linked the peerage and the military establishment.

Another notable military family were the Cathcarts
whose Scottish title dated from 1447 but who had been
given an English earldom in 1814. The 8th, 9th, 10th and
11th Lord Cathcarts had all been distinguished generals
and the military tradition persisted into the nineteenth
century. General Cathcart who was killed at Inkermann in
the Crimea was a son of the 1st Earl, another general.
The 3rd Earl (1828-1905) had joined the army in Canada at
the age of sixteen, though he left it on his marriage in
1850 to take up a public career in county affairs and to
devote his time to 'agricultural business', interests
which brought him the Chairmanship of Quarter Sessions in
the North Riding of Yorkshire for ten years, and the
Presidency of the Royal Agricultural Society. While the
3rd Earl looked after the 5,554 acre family estate, his
brother Augustus (1830-1914) continued to serve in the
Grenadier Guards, eventually retiring as a Lieutenant-
Colonel, while his son the future 4th Earl (1856-1914)
was a Lieutenant in the Scots Guards until he retired in
1881.

This network can also be noted when the relationship
between the officer corps and the baronetage is examined.
The latter, as part of the greater gentry or squirearchy,
formed a very important part of the landed interest, for
in many counties where there was an absence of resident
peers as in Merioneth, or a paucity of them as in the
Welsh counties or in upland areas in England such as
Cumberland, the baronet and his lady were the apex of
local society. Some of this group were landowners on a
large scale. Sir Watkin Williams-Wynn of Denbighshire
(1st Life Guards) owned 145,770 acres, and had an income
of £54,575 a year, an amount which far surpassed that enjoyed by many peers. Others were far less fortunate, but
it was generally assumed, as Burke the genealogist observed in 1861, that a baronet needed an absolute minimum
of landed estate to maintain himself with dignity.(35)

> But for the immense difficulty of rendering, even by
> legislative enactment, real property perpetually in-
> alienable, it might be well that the crown made it a
> condition of conferring an hereditary title that the
> recipient endow the title with a landed estate which
> could never be separated from it. A baronet's qualifi-
> cation might be fixed at £500 a year, a peer's at £2,000.

There are, it must be noted, some examples at the end of
the period to suggest that this landed base was no longer
as important as it had been earlier in the nineteenth century. Thus General Sir Archibald Alison (1826-1907), the
2nd Baronet, who had served in the army from 1846-93, was
typical of a group whose only estate was a London town-
house. Educated at Glasgow and Edinburgh universities,
Alison, whose father, a noted historian, had been created
a Baronet in 1852, enjoyed a distinguished military career
in the Crimea, India and Egypt, but it was evident that he
was no true representative of the landed interest. Nevertheless, in Table 3 where the relationship between the
baronetage and the army is examined, as at 1838, the former may still be closely identified with this landed and
ascriptive base.

TABLE 3   Interpenetration of the army and the baronetage:
1838

Date of creation of title	Number of baronets	Number who held commissioned rank				
		Self	Son	Brother	Kin	Total
Pre. 1660	184	8	6	10	3	27
1661-1770	149	11	4	17	3	35
1701-50	35	1	5	16	7	29
1751-1800	201	10	10	25	4	49
1801-38	281	40	29	48	10	127
Total	850	70	54	116	27	267

When this relationship is looked at more closely, it is
apparent that not only was there a far lower total of officers drawn from this group than from the peerage, but also
that some sections of the baronetage made a low contribution indeed. It is apparent that in the case of the older
titles, it was the minor Scottish and Irish landowners,
raised to the rank of baronet in the seventeenth century,
who provided military officers. This reflected the large
number of Scottish and Irish officers in the Victorian
army. Throughout the period, the officer corps included a
disproportionate number of Irish and Scottish gentry, including a number of clan chiefs. Among these were the two
Cluny Macphersons, Chiefs of the Clan Chattan, MacLaine of
Lochbuie (Murdock Gillan) who was in the 6th Dragoon
Guards, Sir Francis E. MacNaughton (8th Hussars), General

Roderick MacNeill of Barra, Sir James C. MacNeill VC of
Colonsay, Captain Alisdair Robertson (Struan Robertson) of
the Royal Artillery, Alfred Donald MacKintosh (The Mackintosh) of the 71st Highland Regiment of Foot and Sir James
Fergusson of Kilkerran (Grenadier Guards). One of the
most noted of the Irish chieftains was The O'Kelly Mor,
General Sir Richard Kelly (1815-97), whose military career
from 1834 onwards included service in the Crimean War and
the Indian Mutiny. Other officers, however, came either
from families with a long tradition to military service or
from those families in which the head had recently been
created a baronet. In the latter case, the 'military
families' made a contribution to the composition of the
officer corps out of proportion to their numerical
strength. Of the 70 baronets who themselves were army
officers, 25 had been raised to the rank as a reward for
their military service. A further 15 were the sons of
military officers who had been created the first baronet.
In overall terms, these military families contributed 105
officers to the grand total of 267 in this group. It can
also be noted that this pattern was not only exclusive to
the army, for from a total of 138 naval officers who were
similarly related to baronets, no fewer than 90 were members of 29 'naval' families.

When Tables 2 and 3 are totalled, it can be seen that
729 army officers were related either to a peer or a
baronet who was the title holder in 1838. In comparison
with the total number of 6,173 officers, this section of
the landed interest represents nearly 12 per cent of all
serving officers, a percentage which was disproportionate
in relation to the size of this landed interest group.
Nevertheless, this figure is probably an under-estimate of
the actual situation in this year. P.E. Razzell in his
analysis of the social origins of officers during this
period concludes that, '21% of the officers came from the
aristocracy and that an additional 32% came from the
landed gentry as a whole'.(36) This is a more realistic
figure, for it takes into account the position of cadet
families in arriving at a total of 'aristocratic' officers, and examines more closely the contribution made by
those members of the landed gentry who were not baronets.
The possible discrepancy in these percentages is not, however, of importance. The important point is that while
the middle classes were never excluded from membership of
the officer corps, the dominant position of the landed
interest ensured that a military life-style, ethos, norms
and standards were primarily based on the principal
characteristics of the landed interest.

This pattern of recruitment, moreover, remained relatively consistent throughout the period, despite the
organizational changes which were introduced into the army
after 1870, and irrespective of more general social and
political change. The conclusions put forward by Razzell
from a study of three samples taken from the Army Lists of
1830, 1875 and 1912 are summed up in Table 4. (37)

TABLE 4   Socio-economic background of officers in the
British Army

	Aristocracy (%)	Landed gentry (%)	Middle Class (%)
1830	21	32	47
1875	18	32	50
1912	9	32	59

These percentages show the remarkable stability of the
contribution to the officer corps made by the landed gentry between 1830 and 1912. It was not until the Victorian
military establishment had disappeared that officers with
a middle-class background became the majority in the military. The significance of these figures as indicators of
change can, however, be over-emphasized. While it is
evident that, by 1912, three out of every five British
Army officers came from a middle-class environment, this
did not mean that the pattern of recruitment to the officer corps was an 'open' one. Indeed it is clear that
toward the end of the nineteenth century some regiments
became more rather than less exclusive. The aristocracy
maintained its exclusiveness by excluding outsiders from
the 'élite' regiments. In the Household Brigade, for example, the percentage of officers with inherited titles
was higher in 1875 than it had been in 1852. With the
single exception of the Coldstream Guards, the percentage
of aristocratic officers in the élite regiments of the
Household Cavalry and the Brigade of Guards was higher in
1912 than in either 1830 or 1852.(38)
This attempt to maintain the exclusive social status of
the military establishment was common to nearly all regiments in the British Army at this time. To a very large
extent this can be interpreted as a reaction to changes in
the pattern of officer recruitment which were forced upon
a reluctant military establishment by Cardwell's reforms
of 1870-1. But in seeking to maintain control over its
membership, the officer corps was not out of step with
comparable institutions in the rest of society. Here too

it was evident that other occupational groups and organizations were reluctant to yield to external pressure which sought to extend the area from which members could be recruited. This was particularly noticeable in the traditional 'closed' occupations, and the rejection of the spirit of 'open competition' was not unique to the military establishment. A limitation on the area of recruitment was equally evident in other occupations. James Caird in 1878 noted that the professions, Church, army and Civil Service were still all largely recruited from the landed interest.(39) Attempts which had been made by reformers such as Trevelyan and De Fonblanque to widen these sources of recruitment had not met with the success which they had anticipated. The latter, in particular, had recognized that the problem in the military would be particularly difficult to solve. Alone among the reformers who criticized the pattern of officer recruitment, De Fonblanque recognized that the identification of the military as 'aristocratic' did not mean that the army was officered exclusively, or even principally, by members of the nobility:(40)

Such is far from the case. With the exception of the household troops and perhaps a few picked corps, the officers of which belong principally to the titled or untitled nobility, the upper section of the middle class is the one most strongly represented in the higher branches of the army. . . . The term aristocratic, as applied to the constitution of the army, is meant to express that system of exclusiveness, which whether founded upon the test of birth, caste or of money, creates a powerful barrier between the governors and the governed.

So exclusiveness did not depend solely on the rejection by a regiment of non-titled officers. Rather, it resulted from the insistence which was placed on specific ascriptive criteria of selection, that is, criteria primarily associated with membership of the landed interest. Not only were officers from the nobility or greater gentry selected on this basis, but so were other candidates, with the result that the area of possible recruitment into the officer corps continued to be limited. This meant that the majority of officers were recruited from a rural rather than urban society. 'Middle-class officers' often came from what Bateman terms the 'yeomen' (100 to 3,000 acres) or the 'small proprietors' (1 to 100 acres), groups which formed a substantial part of the landed interest. This brought into the army men like William Francis Butler, born in a farmhouse at Ballyslateen, Suirville, Co. Tipperary, the seventh child of a landowner

on a small scale who was always short of money. Nominated to a commission in the 69th Foot at Fermoy by a
distant kinsman, General Sir Richard Doherty, Butler was
almost the epitome of a successful middle-class officer.
Knighted in 1886, a Privy Councillor, a Lieutenant-
General in 1900, he retired to Bansha Castle in Tipperary
after his disagreement with the Colonial Office over
British policy in South Africa before 1900 deprived him
of the military honours which his talents merited.(41)
    Butler was a typical example of the officer who came
from the lesser landed interest. Forced to rely on their
ability to succeed in a military career, this group of
officers often made a substantial contribution to the development of efficiency within the military establishment,
but while there was this very marked identification of
'middle class' with the landed interest, the adopted
selection policy did not exclude entirely other suitable
candidates. Again this helped to prevent the growth of a
military caste, for other middle-class officers came from
professional families. Thus Major-General Sir Owen Burne
(1837-1900), a Director of the Oriental and Peninsular
Steamship Co., was the son of the Reverend H.T. Burne.
Major-General William Byam (1841-1906), son of a surgeon
in Gloucestershire, Captain Hugh Neville (1877-1915), son
of Hugh Nevill of the Ceylon Civil Service, Major-General
John Russell (1839-1900), Extra Equerry to King Edward VII
and son of an Edinburgh lawyer were other examples of the
contribution which a professional middle class, who could
be distinguished from the 'gentry', made to the officer
corps. In addition, a large number of candidates were
themselves sons of officers, but despite the heterogeneous
background of these officers, their education and military
training produced a uniformity of attitudes, expressed in
their willingness to accept a military life-style based on
identification with the landed interest. Deviants from
the norm were exceptional. There were a small number of
officers who were commissioned from the ranks, men like
Colonel Isaac Moore of the 69th Foot, 'an old officer with
the profile of an eagle, the voice of a stentor, and a
heart of great goodness'.(42) Some of these men achieved
high military rank. Major-General Cureton, who was killed
in the Second Sikh War of 1848-9 when the cavalry he was
commanding tried to rescue a gun which had become bogged
down in some river mud, had been commissioned 'for gallant
service' in the Peninsula. Sir Hector Macdonald ('Fighting Mac'), a crofter's son whose life was to end in tra-
gedy when he committed suicide in 1903, served for nine
years in the ranks before he was commissioned after the
Battle of Kandahar. Another Scot, Major-General William

MacBean VC, had won his decoration and a field commission
for gallantry at Lucknow. Major-General Sir Luke O'Connor
VC (1832-1915) had been one of the centre sergeants at
Alma advancing towards the Russian position between the
officers carrying the colours. When one of these was
killed, O'Connor, though himself wounded in the chest,
picked up the colour and carried it to the end of the
action, receiving for this a field commission and the VC.
Most of these officers however, like the Lincolnshire
villager's son William Robertson, who eventually became
Field Marshal Sir William and Chief of the Imperial
General Staff, tended to come from the base of Bateman's
landed interest.

    Among those officers who completely rejected identifi-
cation with the landed interest, General T. Perronet
Thompson was perhaps the extreme example. The son of a
Yorkshire merchant, Thompson was educated at Hull Grammar
School and Queens' College, Cambridge, from which he
graduated as Seventh Wrangler in 1802. Dissuaded from
entering the army, he first joined the Royal Navy in 1803,
but in 1806 he transferred to the 95th Foot. Elected a
Fellow of Queens' in 1804, his academic brilliance brought
him the appointment of Governor of Sierra Leone in 1808
at the age of twenty-five. He was, however, removed from
office in 1810 because of his opposition to the Government's anti-slavery policy. After a turbulent military
career with the Rifle Brigade in Buenos Aires, with the
14th Dragoons in the Peninsula and with the 17th Dragoons
in India, he became Political Agent in the Persian Gulf
before he purchased an unattached lieutenant-colonelcy in
1829 and went onto half pay. Subsequently as owner of the
'Westminster Review' (1820-36) and as a Radical Member
of Parliament he continued a violent anti-Tory campaign,
eventually becoming a strong supporter of the Chartist
Movement. On 9 January 1838, 'The Times' urged that his
name should be removed from the Army List for accusing
the Government of aggression and treason in Canada, but no
action was taken, and in 1868 the inevitable operation of
the seniority rule brought him his ultimate promotion.
Thompson, though, was not the only example of the deviant
officer, and a further two examples are indicative of the
extent to which some officers rejected the life-style
which was expected of them. Edward Mott (1844-1910), the
son of a Staffordshire JP and Deputy-Lieutenant, began his
career in a conventional enough manner. Educated at Eton
and Sandhurst, he was commissioned in the 19th Foot in
1862, serving on the frontiers of India and Burma until
1867 but, a confirmed gambler, he left his regiment to
become a strolling actor before he settled down as

'Nathaniel Gubbins' of the 'Sporting Times'. In contrast,
Harry Panmure Gordon (1837-1902) chose a more conventional
post-military career. Educated at Harrow, Oxford and the
University of Bonn, Gordon, a kinsman of Lord Panmure,
resigned his commission in the 10th Hussars to go to China
where he commanded the Shanghai Mounted Rangers Volunteer
Force during the Taiping Rebellion. Returning to London,
he became a very successful stockbroker, an occupation
which enabled him to follow his hobbies of salmon fishing,
dog breeding and the management of his country estate in
Hertfordshire.

Gordon's life after he entered the London Stock
Exchange is, however, indicative of the extent to which
the acquisition of an estate and the subsequent adoption
of the life-style of the landed interest, was attractive
to a large section of society. Deviants who chose to
reject completely the gentleman's way of life were rare
in number, and very few officers were so committed ideo-
logically that they were prepared to adopt the Radical
attitude which motivated Thompson to attack the entrenched
privileges of a ruling class. The remittance men, that is
the members of the landed interest who chose to 'drop-out'
and avoid the responsibilities they were expected to
assume, were equally rare, and the code of honour which
the officer corps professed to follow often drove them
out of the military establishment. The essential feature
of the officer corps was that while eccentricity was
acceptable, deviancy was not, a philosophy which could be
employed in a wide variety of situations to maintain the
group homogeneity of the military establishment. This was
noted by 'The Times' in 1849.(43)

We understand on undoubted authority that immediately
on the marriage of Lieutenant Heald with the Countess
of Landsfelt (Lola Montez) the Marquis of Londonderry,
Colonel of the 2nd Life Guards, took the most decisive
steps to recommend to Her Majesty that this officer's
resignation of his commission should be insisted upon
and that he should leave the regiment which this unfortunate and extraordinary act may possibly prejudice.

The wish to conform, to be imitative and to identify
with the traditional holders of power was a far from un-
common phenomenon in the officer corps, but it was not
limited to members of the army. The attitude of these
officers reflected a widely prevalent situation in Victorian society. Bulwer Lytton summed this up when he
wrote,(44)

It is the ambition of the rich trader to obtain the
alliance of nobles, and he loves, as well as respects,
those honours to which himself or his children may

aspire. . . . As wealth procures the alliance and
respect of nobles, wealth is affected even when not
possessed; and as fashion which is the creation of an
aristocracy, can only be obtained by resembling the
fashionable; hence each person imitates his fellow,
and hopes to purchase the respectful opinion of others
by renouncing the independence of opinion for himself.
In analysing the relationship between the business and professional men of the 1850s and the land-owning élite,
Aydellotte similarly stresses that stability in England
during this period, in comparison with other European
countries, was largely derived from this wish of an
emerging middle class to be imitative rather than rebel-
lious.(45) What was peculiar to the army was that while
there were signs towards the end of the century that in
the country as a whole these 'hitherto submerged groups'
(46) were beginning to challenge the dominance of the
landed interest, in the military this middle class remained imitative. There are a number of reasons for this,
not least of which was the latent belief that identification with the landed interest ensured stability and the
maintenance of the status quo, thus perpetuating the purpose for which the army existed, but a more immediate
reason was the persistent influence of the military élite.
    The term 'military élite' is a difficult one to define.
It can be interpreted to mean an 'élite by rank', so that
it includes major-generals, lieutenant-generals, and
generals. Alternatively, it can refer to an 'élite by
office', that is, the use of the term recognizes that certain appointments, on the Staff, for example, or as an
Aide-de-Camp, are in some ways 'better' than the more
mundane appointments in day-to-day soldiering. Finally,
it can be used to delineate membership of certain regiments which are believed to be at the centre of military
life, in contrast with those regiments or corps which are
on the periphery of activities. If the first is adopted,
however, it is apparent that the background of senior
officers in the Victorian military establishment changed
very slowly during this period, and it can be inferred
that new entrants to the military establishment, in being
imitative, not only copied the example of an 'external'
landed interest, but also identified themselves with their
own 'internal' élite.
    In specific instances, the narrowness of the base from
which this élite was drawn was most marked. Wellington's
generals, for example, were a remarkably socially homogeneous group. Of the eighty-five officers who served
with him in Spain and at Waterloo, three were peers in
their own right — the 9th Earl of Dalhousie, the 2nd

Earl of Uxbridge and the 5th Baron Aylmer. Fourteen more
were the legitimate sons of peers and one, Beresford, was
the illegitimate son of a Marquess. The fathers of seven
were Baronets, three were judges and one was a Dean in the
Church of Ireland. John Lambert and John Byng came from
naval families. The remainder were mostly the sons of the
greater gentry and the squirearchy, although one, Tilson,
was an early example of the way in which the landed interest and the military adjusted to the pressure of the commercial interest, for Tilson senior was a banker.(47)

More generally, the background of this élite can be
seen in an analysis of the whole group. In Table 5, the
relationship of the military élite to title holders is
analysed for the year 1838.(48)

TABLE 5   Relationship of the military élite to title
holders: 1838

Military	No.	Peer	Son of peer	Baronet	Son of Baronet	Knight	Total
General	91	20	14	12	3	9	58
Lieutenant-General	197	15	15	14	1	50	95
Major-General	219	13	14	7	–	61	95
Total	507	48	43	33	4	120	248

While these figures suggest that a higher proportion of
the military élite than of all officers can be classed as
directly belonging to an aristocratic status group, there
are some limitations on this assumption. The initial
problem is one of status origin. In this table the category of 'peers' includes those officers who had been ele-
vated to the peerage as a reward for their military service, so that it does not necessarily reflect their
original status. Additionally, the majority of officers
at this rank level had a reasonable expectation of receiving a knighthood to confirm their social status, as a
recognition of their loyalty to the establishment. Again
this is no reflection of their original status at birth.
Yet, in excluding cadet members of families, these
figures may be an under-estimate of the true position,
and Razzell concluded that, for 1830, 70 per cent of the
generals came from the aristocracy, 8 per cent from the
landed gentry and only 22 per cent from the middle class.

For major-generals and lieutenant-generals, he found that 57 per cent came from the aristocracy, 32 per cent from the landed gentry and the very small figure of 11 per cent from the middle class. Throughout this period, however, there were variations in the actual number of the peerage who coincidentally held high military rank. In comparison, for example, with the 1838 figures when twenty of the ninety-one generals in the army were peers in their own right, the 1856 Army List names among the seventy-eight generals only eight officers who were themselves hereditary peers: the Earl of Cork, who owned some 38,000 acres in Somerset and Ireland, Lord Seaton and Viscount Gough, both of whom were landowners in Ireland, Earl Beauchamp who controlled 17,000 acres from his seat in Worcestershire, Earl Cathcart of Yorkshire, Lord Downes, the Marquess of Tweeddale, a Scottish landowner, and the Earl of Westmorland, whose military career had been combined with a diplomatic career as Envoy Extraordinary and Minister Plenipotentiary at Berlin. Nevertheless, despite these annual variations, there is a persistent association with the landed interest which increases the significance of these figures, and the precise number of general officers who were peers is not entirely relevant, for this was often an 'accidental' association, in the sense that inheritance of a title from a brother, cousin or more distant kinsman could be fortuitous. Thus Charles Townshend was the heir presumptive of his bachelor cousin, the 6th Marquess, but the marriage of the latter and the birth of an heir apparent ensured that Major-General Charles Townshend did not inherit the family title. Irish and, more particularly, Scottish peerages which could be inherited by 'heirs general' rather than by the more restrictive 'heirs male', often passed through a complex pattern of family relationships, in which a title was unexpectedly inherited by a member of the military élite. The 17th Lord Sempill, William Forbes-Sempill (1836-1905), for example, who had been a junior officer in the Coldstream Guards and who was Colonel of the 4th Battalion, the Gordon Highlanders from 1887, had inherited his Scottish title, created in 1489, and the estates around Craigievar Castle and Fintray House from a kinswoman in 1884. Alternatively, the courtesy title of 'Lord' could be confirmed by the Queen on the younger brother of a peer, in the way in which Lieutenant-General William Seymour (1838-1915), a younger son of Admiral of the Fleet Sir George Seymour, became Lord Seymour in 1871 when his brother became 5th Marquess of Hertford.

What was of much greater significance, was the extent to which these members of the military élite who set the

'tone' of the army and who became a reference group for
their subordinates were members of the landed interest.
Throughout the lifetime of the Victorian military establishment, there were annual variations in the precise
number of these members, as there was in the case of the
military peers, but there was always a persistent association between the military élite and this landed interest.   In 1870 when there were seventy-six generals in the
army, 43 per cent could be identified as the sons of landowners, while in 1897, as the Victorian military establishment was coming to an end, 40 per cent of the generals
and lieutenant-generals were from this group.(49)   As
Razzell comments,(50)

> There are 2½ times as many aristocrats in the ranks of
> Major-General and above as one would expect from the
> proportion of aristocrats in the whole corps. . . .
> This means that the aristocracy maintained their rela-
> tive monopoly of top ranks, although they lost an ab-
> solute monopoly throughout the nineteenth and early
> twentieth centuries.

The influence of these officers permeated the military
establishment.   At one level they were the social leaders
of the officer corps, bringing into the army the attitudes
which had been acquired during their formative years.
Cairnes made this very clear:(51)

> In most districts, the General is the leading light in
> a social sense . . . the presence of any individual
> at the General's dinner table is generally accepted as
> a guarantee of his fitness for any ordinary society.

More importantly, as social leaders, they encouraged the
perpetuation of an amateur tradition, preferring very frequently the recreations of a country gentleman to the responsibilities of soldiering and impressing on their junior
colleagues the need to acquire or develop these recrea-
tional skills.   This was admirably summed up by General
Sir Hew Ross in 1856,(52)

> To render the Artillery efficient, the officers should
> not only be theoretically scientific but they should
> also take an interest in the discipline and instruction
> of the men of their companies;   they should themselves
> be of active habits;   they should be able to control
> and command soldiers;   and encouragement should be
> given to those, who, by their habits of outdoor pur-
> suits acquire the quickness of eye and knowledge of
> horses so indispensable to an Artillery Officer.

This was the translation into the life of the military
establishment of the social philosophy of the landed interest.   It emphasized the paternalism of the latter, so
that the relationship between the officer and his men was

expected to resemble the ideal, if often mythical, relationship between the landlord and his tenant. This was,
at least in theory, dependent on the creation of a trust-
ful and friendly spirit. As Lieutenant-General the Hon.
E.G. French pointed out, the old generals did not ask the
men questions which bristled with hostility to the responsible authority.(53)   Rather, they relied on the existence
of a natural superior-subordinate relationship, of a form
which was epitomized in the comments made by Lady Wantage
on the attitude of her husband, Colonel Lord Wantage
VC:(54)

> This sympathy with the labourers enabled him to gather
> much from them; they were willing to confide in him
> and tell him things, for as one of them remarked, 'He
> does not talk to us as many gentlemen do, of things
> we do not understand; he seems like a friend like.'
> He was frequently accompanied in his walks by some old
> labourer whose mature intelligence and practical ex-
> perience he appreciated; and the talk was profitable
> and instructive to both.

The legacy of this paternalism was hard to eradicate.
Even Wolseley who, as an ambitious young colonel had
attacked many military myths in 'The Soldier's Pocket
Book', published in 1869, including a denunciation of the
phrase 'officer and gentleman' for which he wanted to substitute the simple word 'soldier', seemed to condone the
existence of this attitude. His criticism was directed
at officers who kept order by a rigid system of espionage
and who had but little real sympathy between them and
their men. What was needed, it appeared, was officers
who knew their men, who were able to exercise the meticu-
lous direction and quasi-paternal oversight over subordinates which was considered indispensable in the management of an estate or in the control of a regiment. The
essentials of the attitude needed to fulfil this role was
summed up in a vivid description by Steevens of the
British Officer on the banks of the Atbara during the reconquest of the Sudan (1895-9):(55)

> When he got up in the morning he had nothing to shave
> with, and lucky if he got a wash. The one camel-load
> of mess stores was well nigh eaten up by now; he
> received the same ration as the men. His one shirt
> was no longer clean; he hardly dared pull out his one
> handkerchief; he went barefoot inside his boots while
> his socks were being washed. And always — night or
> day, on fatigue or at leisure, relatively clean or
> unredeemedly dirty, when he had borrowed a shave and
> felt almost like a gentleman again, or when he lay with
> his head in the dust and the black private doubted

whether he should salute or not — his first paternal
thought was the well being of his men.

Steevens, perhaps, was over-sympathetic in his description, but the contrast with the comments made by critics
of the officer corps in the early years of the Victorian
army cannot be simply dismissed. Changes had occurred in
the attitudes and behaviour of officers during this
period, but the common standard of expected behaviour was
admirably summed up by General Sir George Higginson (1826-
1915), who had served with the Grenadier Guards for
thirty years:(56)

Our commanding officers were men of the world,
thoroughly conversant with social and political life;
if they preferred life in a wider community to the
somewhat narrow sphere of regimental command, they
took the most high minded view of the work of their
subordinates, holding themselves absolutely responsible for any error of judgement which they might
betray and were generous in their praise of duties
well performed.

In encouraging the development of this behavioural
standard, officers who came from the landed interest were
aided by those members of the military élite who came
from families with a tradition of service in the army.
In 1897, almost half of the sixty-three generals and
lieutenant-generals came from military families. Some
of these also belonged to the landed interest, but it is
significant that 38 per cent were from propertyless
families and to a certain extent this does qualify the
earlier assertion of a closed and privileged élite. At
the same time, however, these were men born and raised in
the military tradition, who valued the ethos and norms of
the Victorian military establishment, and who frequently
through marriage identified themselves with the traditional landed interest. In 1829, Henry George Hart, for
example, was nominated to a commission without purchase in
the 49th Foot. Nineteen years later he purchased his
majority, became Depot Major at Templemore, the Irish
home of the 13th, 41st, 47th and 55th Foot, and eventually
reached the rank of Lieutenant-General. His son, Arthur
Fitzroy (1844-1910), who was educated in a conventional
middle-class manner at Cheltenham and Sandhurst, joined
the 31st Foot in 1864. Service in the Ashanti War, the
Zulu War, the 1881 Boer War and the Egyptian War of 1882
where he was Wolseley's Deputy Assistant Adjutant-
General, culminated in the command of the Irish Brigade in
South Africa and promotion to Major-General. By this
time, however, Hart had considerable connections with the
landed interest, for in 1868 he had married May Synnot,

daughter of Mark Synnot of Armagh, where he owned 7,321 acres at Ballymoyer. When his wife eventually succeeded to the estate, Hart assumed by Royal Licence the additional surname and arms quarterly of Synnot, and in his retirement he lived the life of an established member of the landed interest as an Irish landowner and JP.

Frequently, these marriages produced a complex network of interrelationships with the landed interest. General Sir Robert Gardiner, for example, joined the Royal Artillery in 1797, serving in the Peninsula campaign and at Waterloo. His connection with both the military and landed interest was confirmed when he married Caroline, daughter of General Sir John Macleod, Colonel-Commandant of the Royal Artillery and Lady Emily, daughter of the 4th Marquess of Lothian. This link was furthered by the son, General Sir (Henry) Lynedoch Gardiner (1800-97) who, having been educated at Cheam School and the Royal Military Academy Woolwich, joined the Royal Artillery in 1837 at the age of seventeen. After service in Canada, in the Crimea and the Indian Mutiny, Sir Lynedoch was from 1872-96 Equerry to Queen Victoria, and he then became Bath King of Arms. The son of a General, the grandson of a General and the great-grandson of General the 4th Marquess of Lothian who had fought at Fontenoy and who commanded the cavalry on the Royalist left at Culloden, Gardiner confirmed his membership of the military and landed interest when in 1849 as a young artillery captain he married Frances, daughter of Francis Newdigate and Lady Barbara, daughter of the 3rd Earl of Dartmouth. Sir Henry, who, like his father, had been knighted by the Queen, was thus connected with both Scottish and English landed interests, though he and his wife lived in a royal grace and favour residence in Richmond Park, and lacked an estate of their own.

It was these links with the landed interest which drew attention again to the way in which comparable situations in England and Germany in the nineteenth century, where precisely the same average of the members of each officer corps were sons of officers, produced such disparate attitudes.(57) The militaristic attributes of the German Junker class were not reproduced in England, where most members of the military élite saw themselves as gentlemen first and officers second. The interpenetration of the civil and military élites modified considerably any tendency on the part of the army towards professional insularity. There are a large number of examples of this. General Sir Rufus Donkin was himself the son of a General but his army career from 1778 onwards, including service in the Peninsula, did not hinder his participation in

politics as member in the House of Commons for Berwick
from 1832-7 and Sandwich in 1839 or as Governor of the
Cape of Good Hope. General Sir W.H. Clinton and
Lieutenant-General Sir H. Clinton followed their father
into the army but combined this with the management of
their estates and membership of the House of Commons.
J.T. Clifton and his son T.H. Clifton both served in the
1st Life Guards before retiring to their Lancashire
estate to play their part in local administration and to
represent North Lancashire in the House of Commons as the
Conservative member from 1844 to 1847, and from 1874-80
respectively. In every county of England, and more particularly in Scotland and Ireland, there were examples of
these military families who combined a tradition of army
service with a persistent involvement in national and
local affairs. Their attitude towards 'service' was
furthered, moreover, by the way in which recruitment from
other occupational groups was regulated on the basis of a
latent if not a manifest identification with this traditional ethos. Thus while one quarter of the military
élite in 1897 had been recruited from a middle-class professional background, only 5 per cent came from the commercial interest. Moreover many of these professionals
overlapped with the landed interest, sharing a common
background and a mutual love of the landed way of life.
Thus, the father of General Sir Michael Biddulph (1823-
1904), Gentleman Usher of the Black Rod from 1896, was
the Reverend T.S. Biddulph of Amroth Castle in Pembrokeshire, but the General, although he lived in a London
flat, followed the life-style of the landed interest with
his commitment to hunting, shooting, sailing and hawking,
though his fondness for sketching may have distinguished
him from more philistine landowners. Similarly, General
Sir James Browne (1823-1911), who had 'worn the Queen's
uniform boy and man from 1838 and had served in the four
quarters of the globe'(58) and who was the son of the
Honourable the Dean of Lismore, boasted of the fact that
his love of all field sports had led him to hunt the
buffalo and the wapiti on the plains of North America and
the moose in Nova Scotia, even though after his retirement
from the post of Governor of the Royal Military Academy
in 1887 he lived in South Kensington.
   To the lay critic, it often appeared as though this
military élite was the last bastion of neo-feudal characteristics and attitudes. A contemporary comment alleged
that 'The Upper classes betake themselves to the army
for exercise, companionship and enjoyment. War is the
occupation of the nobility and the gentry',(59) and while
this comment was applicable to the whole of the officer

corps, it summed up, very clearly, general conclusions
about the characteristics of these military leaders.
This comment, too, seemed to be evidenced by the way
in which when appointments were made from this élite to
command appointments, selection appeared to prefer the
officer from the nobility and the gentry to his middle-
class colleagues. In Scotland in 1856, for example, the
General Officer Commanding was Major-General Viscount
Melville of Melville Castle, Edinburgh, and the Assistant Adjutant-General, Colonel Hon. A.A. Dalzell, kinsman of the Earl of Carnwath. In Ireland, General Lord
Seaton's Military Secretary was his son Major Hon. James
Colborne, his Aides-de-Camp Captain Sir Lydston Newman
(7th Hussars), owner of over 5,000 acres in Devonshire,
and Major Hon. F.J. Evans Freke of the 2nd Life Guards.
In India, the Staff List read like a page from Burke's
'Peerage' and 'Landed Gentry', from the Commander-in-
Chief, Lieutenant-General Hon. George Anson, brother of
the Earl of Lichfield, to his Aides, Lieutenant-Colonel
Charles Denison of the 52nd Foot and Captain Hon. E.J.W.
Forester of the 83rd Foot.
    It was the persistence of this apparent inter-
relationship between the military élite and the landed
interest, which seemed to typify the contribution made
by the peerage and gentry to the Victorian army. The
Denison family, for example, were typical of the way in
which members of the landed interest permeated the military and other institutions in society. Of the nine sons
and five daughters of John Denison of Ossington Hall,
Newark, an estate of 6,000 acres, two sons entered the
army. William went from Eton to Woolwich, and then into
the Royal Engineers. A Lieutenant in 1830, he followed
up an army career by becoming Governor of Van Dieman's
Land in 1846, and Governor of New South Wales in 1846 as
Lieutenant-Colonel Sir William Denison. Charles, who
joined the 52nd (Oxfordshire) Foot in 1837, and later became its Colonel, had a number of staff appointments, including that of ADC to Lieutenant-General Hon. George
Anson. These brought him very much into the Indian public
service and 'till he was compelled by family health,
caused by sun-stroke, to retire from active life, was
Chief Commissioner of Civil Service at Madras'. The
eldest brother, John, following the father into Parliament became Speaker of the House of Commons, and on retirement was created Viscount Ossington. Another brother,
Edward, was Bishop of Salisbury. Henry, who like his
brothers had gone up to Oxford, stayed as a Fellow of All
Souls, while Stephen became Deputy Judge Advocate. Frank,
described as 'the sailor' died in 1841. Alfred 'after

twenty years of laborious, honourable and successful life
in Australia' became Private Secretary to the Speaker. The
ninth brother, G.A., was a clergyman with a brilliant
academic reford, who became very involved in questions of
educational reform.(60)

Statistically, a large part of the officer corps at
any one time could be described as of middle-class origins, but statistical facts were of less relevance than
the feeling, commonly accepted in the United Kingdom, that
the military was, in De Fonblanque's terms, 'aristocratic'.
Throughout the nineteenth century the military was seen
as a way of life, an occupation with few material rewards
attached to it, but conferring a number of advantages on
its new members. In a deferential élitist society, where
'The English of all ranks and classes, are at bottom, in
all their feelings, aristocrats',(61) the parent society
wanted to believe that its army could be identified with
the landed interest. The virtues of the former —
bravery, discipline, obedience, absence of reward, and
patriotism — were equated with the merits of the latter.
For the aspirant to the military way of life, the perpetuation of these virtues encouraged his identification
with the life-style of the landed interest, and society
accepted the need for this imitation, for(62)

> The exercise of these virtues to a very high degree is
> so essential to the career of arms that they consti-
> tute its characteristic feature, define its own pecu-
> liar spirit. They are the necessary conditions of the
> existence of the career, and if they disappeared it
> would disappear also.

So, to the majority of the parent society, the Victorian
army projected an image which was accepted as essentially
aristocratic. Officers, like Members of Parliament,
appeared to have(63)

> a common freemasonry of blood, a common education,
> common pursuits, common ideas, a common dialect, a
> common religion and — what more than any other thing
> binds men together — a common prestige, a prestige
> growled at occasionally, but on the whole conceded,
> and even, it must be owned, secretly liked by the
> country at large.

3

⋄⋄⋄

The Purchase System

⋄⋄⋄

While a large part of Victorian society were eager to
identify their military with the aristocracy and the
landed interest, a smaller section were quick to criticize the apparent effects of a restricted policy of recruitment. For some of these critics, their evaluation
of the army was based on personal experience. To
Radicals such as William Cobbett, the military was far
from superior or gentlemanly:(1)

> Those who were commanding me to move my hands or my
> feet thus, or thus, were in fact, uttering words which
> I had taught them; and were, in everything except
> mere authority, my inferiors; and ought to have been
> commanded by me. . . . But I had a very delicate part
> to act with these gentlemen, for while I despised
> them for their gross ignorance and their vanity, and
> hated them for their drunkenness and their rapacity,
> I was fully sensible of their power.

A less emotional evaluation of the value of the contribution made by the landed interest to the military establishment came from reformers such as Trevelyan and
De Fonblanque. These critics were primarily concerned
with the effect of patronage in the army on military professionalism. To them the crux of the problems faced by
the Victorian army, problems which had become very evident
in the Crimean débâcle, was the existence of the purchase
system. They did not ignore the effect of the system on
the military as an organization. They argued vehemently
against the contention that the system had to be kept in
being if the military were to be an apolitical force. But
primarily their criticism of the purchase system was based
on the presumed effect which it had on the development of
military professionalism. To Trevelyan, it was evident
that(2)

The large and important class of well-educated young
men who depend for their advancement upon their own
exertions, and not upon their wealth and connections
and who constitute the pith of the Law, the Church,
the Indian Civil Service and other active professions,
are thus ordinarily excluded from the Army.

From this it was deduced that while the country as a
whole in the years following the Crimean scandal wanted
to raise the army in the scale of professions, 'these
natural and wholesome aspirations are repelled by the
purchase system'.(3)

So, the fundamental question under discussion was
whether the officer needed to be an amateur, 'a real
gentleman, a truly noble man, a man worthy to command, a
disinterested man of integrity, capable of exposing, even
sacrificing himself for those he leads',(4) or a professional 'whose only claim to a commission had to be in
time of peace, knowledge and education; in war, courage
and conduct'.(5) The problem which persisted, was that
as long as entry into the military establishment was
governed by a system which allowed a man to purchase his
commission and his promotion, it seemed inevitable that
the amateur with money would rise to the top. Professionalism, it was alleged, was not dependent on the limited area of recruitment, but on the system of recruitment:(6)

Purchase and professional qualifications are antagonis-
tic and incompatible principles. We must take our
choice of them. The army cannot be constituted upon
both at the same time. If certain sums of money have
to be deposited as the condition of successive steps
of appointment and promotion, those who have ready
money at their disposal must be appointed and promoted
even in preference to other better qualified persons.

The question which arose was the extent to which it was
the existence of the purchase system, rather than the
limited area of recruitment into the Victorian army,
which affected the development of military professionalism.
To this, there was no ready answer. Supporters of patronage, which apparently depended for its existence on the
continuance of the purchase system, could still argue that
it had brought into the army men of distinction. Advo-
cates of competition could equally argue that the Crimean
War had clearly demonstrated its faults. To look further
for an answer, it is necessary to examine in more detail
the characteristics and the nature of the purchase system
itself.

To Max Weber, the system of purchasing commissions in
the British Army was a form of appropriation, a charac-

teristic of traditional patrimonial authority. It suggested that particular powers and corresponding economic
advantages had become the monopoly of a specific social
group, the members of which enjoyed a comparatively well-
defined status with a common mode of life and a detailed
code of behaviour.(7)   It was this 'monopoly' which the
spokesmen for a growing middle class were now challenging.
To them the system appeared to be an outmoded relic of
feudalism which was incompatible with the development of
a rational, legal, bureaucratic society in an era of technological development and social change. Historically, in
Great Britain the system was post-medieval in origin,
since it had developed in the seventeenth century when it
was generally accepted that civil offices under the Crown
could be bought and sold,(8) but its characteristics to
these critics were still tainted by neo-feudalism. It
suggested that officers continued to look upon themselves
as soldier adventurers who had organized themselves into
companies for hire by any employer who needed the support
of a military force. In these organizations the underlying principle which had motivated recruits was the hope
of prize money, private or corporate, the latter being
distributed on a fixed scale according to the rank of
every officer. Thus each rank had come to possess a
pecuniary value, for shares in potential corporate prize
money could be sold like shares in any other commercial
enterprise. So to the critics of the system, it seemed
that Victorian commissions were still sold to 'adventurers', albeit that the prize to be shared was no longer
booty or money as such, but a less tangible form of
advantage.
      The system in the nineteenth century was peculiar to
the British military, and, despite the criticisms which
were raised against the system by those reformers who
favoured a meritocratic military, it was supported by the
majority of the officer corps, so that there were very
few complaints from individuals themselves. This was so,
even though a certain amount of bitterness might have been
expected from officers who lacked the money to take advantage of the way in which the system operated. But such
bitterness was rarely evident and most existing members
of the military and aspirants to the officer corps were
content to accept that this was a part of army life which
had a lengthy historical background. The latter was evidenced by the way in which, originally, the system had
recognized the custom of granting commissions to gentlemen for their having raised a body of troops in the King's
Service. The tradition that the regiments should be
equipped at the expense of their officers and that they

would be generally raised in the district where the
colonel's estates lay, was very marked in the Civil
War.(9)   As Clarendon points out,(10)

> When the King had settled his Court at Oxford . . .
> he gave Banbury . . . to the Earl of Northampton who
> was commissioned to raise a regiment of Horse, which
> was given to Lord Compton, his eldest son, and Sir
> Charles, his second son, was made Lieutenant-Colonel
> of it.

In part this too was an historical legacy, for it was derived from the chivalric and feudal concept of 'Tenure by
Knight Service' although this was formally abolished in
the Tenures Abolition Act of 1660. Nevertheless, after
this date, a colonel who raised a body of troops for the
King still had the trouble and expense of the recruiting
campaign in which he was forced to stand certain costs
from his own pocket. To recoup himself, the colonel expected new recruits, whom he could nominate to specific
ranks in the regiment, to contribute towards those expenses; in turn, these officers expected to recoup their
initial expenditure by selling their commission to a
successor.(11)

In the Victorian military establishment there were,
until the abolition of the system in 1871, two main problem areas. First, there was considerable discussion
about the validity of an inherited system which, in per-
mitting the purchase of the initial appointment, seemed
to favour the entrant with money rather than the entrant
with ability. Second, a more controversial dispute
centred around a system in which promotion was purchased
rather than awarded on the basis of merit. While both
systems were amended during this period to meet external
pressures, the general principles in each case remained
constant. Let us look at each of these in turn.

The entrant, who had to be sixteen, first obtained a
nomination from either the Commander-in-Chief at Horse
Guards, or, if he wanted to enter the Household Regiments
of Horse and Foot Guards, from the colonel of his selected regiment. If approved, his name was placed on the
list of those eligible for commissioning and he waited
for a vacancy to arise in the regiment of his choice. In
the absence of 'interest', which kept his name to the
fore, it was suggested that he could wait for a considerable length of time.(12) For a number of candidates this
was a period during which they avidly canvassed supporters
to ensure that their names were not overlooked. Eventually, if an officer in the selected regiment decided to
sell out, either by leaving the army or by obtaining
promotion thereby causing a vacancy, the candidate's

commission was gazetted and he paid the appropriate purchase price.

TABLE 6   Regulation prices of commissions: 1821 Warrant

Commissions	Prices	Difference in value between the several commissions in succession
	Royal Horse Guards	
	£	£
Cornet	1,200	-
Lieutenant	1,600	400
Captain	3,500	1,900
Major	5,350	1,850
Lieutenant-Colonel	7,250	1,900
	Life Guards	
	£	£
Cornet	1,260	-
Lieutenant	1,785	525
Captain	3,500	1,715
Major	5,350	1,850
Lieutenant-Colonel	7,250	1,900
	Dragoon Guards and Dragoons	
	£	£
Cornet	840	-
Lieutenant	1,190	350
Captain	3,225	2,035
Major	4,575	1,350
Lieutenant-Colonel	6,175	1,600

Commissions	Prices	Difference in value between the several commissions in succession
Foot Guards		
	£	£
Ensign	1,200	–
Lieutenant	2,050	850
Captain with rank of Lieutenant-Colonel	4,800	2,750
Major with rank of Colonel	8,300	3,500
Lieutenant-Colonel	9,000	700
Marching Regiments of Foot		
	£	£
Ensign	450	–
Lieutenant	700	250
Captain	1,800	1,100
Major	3,200	1,400
Lieutenant-Colonel	4,500	1,300

This price had originally been laid down by a Royal Warrant of 10 February 1766, but the tariff of regulation prices was slightly altered in 1783 for cavalry regiments, and in 1821 was further amended for the Guards and the Line Regiments. The 1821 Warrant, reproduced as Table 6, was an attempt to regularize a known situation in which over-regulation prices had been paid, 'with a view of adapting the same to the present circumstances, and the general interest of the Service'.(13) Nevertheless, although the payment of these extra amounts was specifically forbidden in the Royal Warrant and although it was contrary to civil law,(14) further increased prices were continually demanded and paid after 1821.

The purchase of a commission was not the only method of entry into the military establishment, and although public criticism of the system tended to overlook this point, a

certain number of non-purchase commissions were available
each year. Entrants into the Engineers and Artillery,
that is, the Scientific Corps, who underwent a course of
training at the Royal Military Academy, Woolwich, did not
purchase a commission. Neither after 1842 did those entrants who entered the army via the Royal Military College
at Sandhurst. Additionally, even before this date a number of direct-entry commissions were available without
purchase, while a small number of officers, as is shown in
Table 7,(15) were commissioned from the ranks.

TABLE 7 First commissions granted in the army: 1834-8

Regiments	Commission by purchase	Commission by non-purchase	Total
Cavalry	221	6	227
Guards	34	8	42
Line	859	246	1,105
Total	1,114	260	1,374
From the ranks	3	33	36
Total	1,117	293	1,410

The awarding of non-purchase commissions was, it was
alleged, not designed to recruit into the military entrants of ability, for 'in no case, are the personal
merits of the applicant considered for one single moment'.(16) These commissions were still at the nomination
of the Commander-in-Chief, so that, primarily, this was an
extension of his powers of patronage. At the same time,
this did prevent domination of the officer corps by men
with wealth alone, and it did enable the Commander-in-
Chief to bring into the military potential soldiers, such
as Garnet Wolseley, commissioned without purchase as an
Ensign on 12 March 1852, who otherwise might have been
completely excluded from the establishment. A feature
of this policy, moreover, was the extent to which free
commissions were granted to the sons of officers, either
directly to a regiment or by nomination to Sandhurst.
Among the first officers to be commissioned without purchase at Sandhurst was Frederick Middleton, subsequently
Lieutenant-General Sir Frederick (1825-98). The son of
Major-General C. Middleton, Frederick, who was educated at
Maidstone Grammar School before he joined the Royal Military College was commissioned as an Ensign in the 58th

Foot in 1842. After seeing active service in New Zealand,
the Santhal Rebellion, and the Indian Mutiny, he passed
through the Staff College in 1865-6 and became Commandant
of Sandhurst from 1874-84. The following year he took
command of the successful expedition against the Riel
Rising in the North-West of Canada, at the conclusion of
which he received from Queen Victoria a knighthood and
promotion to the rank of Major-General, from the Canadian
Government a grant of £4,000 and from the War Office a
pension of £100 a year. After retiring from Canada in
1890, he was appointed Keeper of the Crown Jewels in the
Tower of London. Another of the non-purchase Sandhurst
officers in these early days was Lieutenant-Colonel Sir
John T. O'Brien (1830-1903), son of Major-General T.
O'Brien, acting Governor of Ceylon. After service with
the Indian Army, Sir John was employed in a number of
civil posts including those of Inspector-General of
Police, Mauritius, Governor of Heligoland, and finally
Governor of Newfoundland. These officers were examples
of the many more whose commissioning generated deliberately or unconsciously a pattern of self-recruitment which
developed as a counter-balance to the interest group of
aristocratic officers who might have joined the army for
a limited period. Similarly, the Commander-in-Chief, by
a judicious use of patronage, could 'control' the balance
in the intake of aspirants by regulating the number of
free commissions which were available. It could also be
suggested that while the Commander-in-Chief still deliberately ensured the social cohesiveness of the officer
corps, a latent effect of his right of nomination to a
commission without purchase was to create a sub-group of
officers who were, from the beginning, indebted to him.
   The majority of these commissions were usually granted
in the less exclusive line regiments, and in peacetime
few free commissions were granted in the Cavalry, other
than to selected soldiers who were commissioned as
'riding-masters'. Indeed, in the Guards, the eight non-
purchase commissions listed in Table 7 were awarded to the
Sovereign's Pages of Honour, on the basis of two per year.
At the same time, during a war such as the Crimean War,
a considerable number of free commissions were awarded.
Not only was this necessary to meet the needs of wartime
expansion and casualties, but there was also a continuing
need to provide an incentive for recruitment, not least
of all because the value of a purchased commission was
until 1856 lost on the death of the holder.

TABLE 8 First commissions granted in the army: 1855

Regiments	Commissions by purchase	Commissions by non-purchase	Total
Cavalry	31	44	75
Guards	13	16	29
Line	258	949	1207
Rifle Brigade	13	54	67
Total	315	1063	1378

   Behind these figures(17) can be seen some of the effects of the Crimean War. In the cavalry, for example,
in which few non-purchase commissions had hitherto been
granted, the majority of commissions gazetted in 1855 were
free, and in some regiments, such as the 13th Light
Dragoons, which had suffered severely in the Charge of the
Light Brigade, a relatively high number of these commissions was awarded. In the Line regiments, there were considerable differences in the policy adopted during this
year. In the 75th Foot, stationed in Bengal, there were
no regimental vacancies after Charles Pym purchased his
commission in 1853, so that the question of awarding free
commissions did not arise. In contrast, in the 34th
(The Cumberland) Foot serving in the Crimea twenty-three
non-purchase commissions were gazetted in 1855. For
these entrants promotion could be rapid, and less than
eight months after commissioning Lieutenants Noel Harris
and Julius Laurie of the 34th were severely wounded in
the storming of the Redan on 8 September. Other battles
in the Crimea similarly took their toll of junior officers. At Alma, the 1st (The Royal Fusiliers), the 23rd
(The Royal Welch Fusiliers) and the 33rd (The Duke of
Wellington's) Regiments of Foot were severely mauled by
the Russian artillery as they made a frontal attack on
the Great Redoubt. In these three regiments alone, to
replace casualties, seventy-three non-purchase commissions
and four by purchase were gazetted between January and
November 1855.
   This number of free commissions was exceptional and
when peacetime returned the traditional pattern of purchase and promotion was re-instituted, although at one
time sixty-five cornetcies which were available in the
cavalry regiments could not be filled because of a lack
of applicants. The purchase of the initial commission
continued to attract considerable criticism. 'It was',

argued Lord Stanley in the House of Commons during a debate on a motion to establish a Select Committee to consider the question, 'the buying by a private person of a
vested interest in the public service. . . . In the British
Army we had this system which was not accepted by any
other nation, which was condemned by the example of the
civil service, and which was not found in the navy. It
was essentially anomalous and exceptional.'(18) Equally,
it had its defenders who continued to argue that the system encouraged the recruitment of officers who would be
independent of spirit and 'secure against the influence
of favour'.(19) It was, however, the practice of purchasing promotion which attracted the greater criticism
for, here, certain clear advantages appeared to be given
to the man with money who could buy over the heads of
more experienced and better-qualified poor officers. It
was also argued that the system discouraged initiative,
since 'it deadens the feelings of emulation and the eagerness to acquire military knowledge, and it renders men
eligible for the highest command without taking any
security that they are fitted for such a position'.(20)
This criticism could be attributed to the failure until
1850 to test in his knowledge of military affairs the
officer who sought promotion, and to the practice of promoting up to, and including the rank of, lieutenant-
colonel solely on the basis of seniority. Thus when a
vacancy arose within a regiment, it was offered in turn
to all officers who were qualified to 'buy' the next step,
with the proviso that they lost their claim to purchase
the promotion if an objection had been lodged against
them by their commanding officer or the Commander-in-
Chief.(21) Periodical returns were made from every regiment to the Commander-in-Chief showing the names of officers of each rank who wanted to purchase advancement and
the references which they gave for the funds necessary to
effect this. The commanding officer of the regiment
certified that the applicants were 'fit and qualified' for
promotion, and two principles governed the subsequent
purchase of promotion. First it was laid down that no
officer, however deserving, would be promoted without
purchase over the head of his senior in the service;
second, it was a mandatory rule that no officer would be
promoted by purchase over the head of his senior officer
in the regiment, provided that this senior officer had
stated his claim to purchase according to the regulations.
     In theory, these rules meant that when, for example, a
vacancy arose for a major in a line regiment it was initially offered to the senior eligible captain at a price
of £3,200. If he accepted, he could sell his captain's

commission for £1,800 and his promotion had cost him the
'difference in value between the several commissions'.
If the senior captain did not want to purchase the majority, then the other captains were offered the vacancy in
order of their seniority. If no qualified officer of the
regiment presented himself for purchase, then an officer
was brought in from another regiment to fill the vacancy
or was transferred from the half pay list. In practice,
however, the operation of market forces, particularly in
the more popular and fashionable regiments, encouraged the
payment of sums beyond the regulation price with the result that the latter had become the minimum rather than
the maximum charge.(22) It was this abuse of the system
which generated the most violent criticism of the system,
for to the layman this 'gross, widespread and mischievous
illegality', as it was termed by Gladstone, turned the
military establishment into a collection of shopkeepers.
Nor were officers themselves less ready to object to
it,(23) for it was, as Colonel Lord West told the Purchase
Commission, a degrading and sordid traffic.(24) The
offence of over-payment was one for which the 'Manual of
Military Law' prescribed cashiering, whilst it was equally
an offence under civil law, but the system flourished and
little apparently could be done to prevent it.(25) Occasionally the law was apparently satisfied through the
payment by the purchaser to the vendor of an excessive
amount for a horse or other item for equipment whilst coincidentally the commission changed hands at the regulation price. More usually, however, the over-regulation
price was simply asked for and paid.

     The amounts involved are not entirely certain. Cecil
Woodham-Smith, quoting from 'The Times', says that in 1832
Lord Brudenell (later Earl of Cardigan) bought the
Lieutenant-Colonelcy of the 15th Hussars, a fashionable
élite regiment, for between £35,000 and £40,000 against a
regulation price of £6,175. Coincidentally Brudenell
passed over the head of the senior Major, Anthony Bacon,
who sold out and entered the Portuguese Army.(26)
Christopher Hibbert, citing 'The Letters of Captain
C.M.J.D. Shakespear RHA', suggests that on one occasion
£57,000 was paid for a lieutenant-colonelcy.(27) Charles
Hammersley in his evidence to the Purchase Commission of
1856 stated that he had known £14,000 to be the common
price for the lieutenant-colonelcy of a cavalry regiment
and as much as £13,200 to be given for the command of a
guards regiment. Similarly the price commonly given for
a lieutenant-colonelcy of a line regiment was £7,000 in
comparison with the regulation price of £4,500 and £2,400
was paid for a captaincy against the regulation price of

£1,800.(28)   Autobiographies of the military élite sometimes refer to their own personal experiences. Sir Evelyn
Wood (1838-1911), who began his career in 1852 as a midshipman in the navy, serving with the Naval Brigade in the
Crimea, was quite frank about his experiences of purchasing
promotion after he had transferred to the army. In 'From
Midshipman to Field Marshal' (1906) he claims that in 1861
when he became a captain in the 17th Lancers, he paid
£1,000 to the Government and £1,500 over the regulation
price to the officer who retired in his favour. Ten years
later, shortly before the purchase system was abolished,
he was negotiating with the majors of three regiments,
arranging to pay various sums from £1,500 to £2,000 for an
exchange.

This trafficking in promotion was particularly noticeable at the apex of the regimental rank structure. The
lieutenant-colonel of a regiment who after three years'
service in the rank automatically received the army rank
of colonel was faced with a considerable problem. His
promotion to the army rank of major-general was guaranteed
through the operation of the seniority rule, but promotion
also meant that he could not sell his regimental commission. The justification for this rule was derived from a
belief that if officers at this rank level knew they
would receive the value of their commissions at any time
and without any risk, they would never retire. This
would not only have blocked the promotion of younger officers, but it would have also resulted in the age of
officers steadily rising until the regiments were full of
superannuated individuals. But faced with the possibility
of losing their substantial investment in the system officers were faced with a very real difficulty as their promotion to major-general became imminent. The regulations
were rigidly enforced, so that their simple choice, in
theory, was promotion or retirement. In practice, however,
a complex set of unofficial customs had developed. Since
transfer to the half pay list did not affect promotion,
the officer could adopt this solution, although by the
1850s voluntary transfer to half pay was limited to those
officers who were either medically unfit or who had served
for at least twenty-one years.(29) More importantly, once
on half pay an officer could not sell his commission. The
way around this was spelt out by Colonel Lord West in his
evidence to the Purchase Commission:(30)

> I will take the stance of a lieutenant-colonel who has
> paid largely above the regulation price for his com-
> mission and who wishes to retire on half-pay. Of
> course he wishes to get back a portion of that large
> sum which he has paid down, and what are the steps that

he takes to effect this? He first has to cast about,
and he probably employs one of the numerous army agents
to find an officer already on the half-pay list, and
who, for the consideration of a few hundreds, will
exchange with him. Then the same lieutenant-colonel
says to his officers 'How much will you give me if I
manage this transaction?' The officers make up a
purse for him, and the required sum also above the
regulation for the lieutenant-colonel on half-pay who
takes his place on full pay in the regiment in order to
sell out, but in order to carry out this arrangement,
a great deal of manoeuvering and a great deal of
trafficking is necessary.

This solution, as in France during the eighteenth
century where a similar system known as the Concordat had
governed the sale of captaincies and lieutenant-colonel-
cies,(31) satisfied all the parties concerned. The
lieutenant-colonel, now on half pay, continued to be
eligible for promotion to the rank of major-general whilst
he had also recovered part of the cost of his commission.
The former half-pay officer had been able to commute his
half pay into a capital sum and although he had surrendered
his half pay on returning to the active list prior to retiring by selling his commission, this was not necessarily
disadvantageous. Many of these 'exchange' officers had
received their promotion on the half-pay list but this had
not increased the amount of money which they received, for
this was based on the pay of their rank when they left the
active list, so that frequently there was a financial advantage to be gained through the sale of the exchanged
commission. Equally, the officers involved in the Concordat were satisfied since the sale of the lieutenant-
colonelcy to the senior major created subordinate regimental vacancies.

Regimental officers, however, could and often did
refuse to adopt this course of action, hoping that the
vacancy would inevitably arise on the promotion of their
lieutenant-colonel. They took some risks, however, for
the thwarted officer could exchange with a lieutenant-
colonel in another regiment and then seek to come to an
arrangement with the officers in his new regiment. Even
if this did not happen, there was no guarantee that the
vacancy which 'fell' into the Horse Guards, would not be
used by the Commander-in-Chief to reward a distinguished
major by transferring him to the vacant lieutenant-
colonelcy. Occasionally, a colonel could find himself
'caught in a brevet' if a sudden increase in the number
of promotions to major-general, as occurred during the
Crimean War, caught him unawares, although usually the

inevitability of promotion to general officer rank through the rigid operation of the seniority rule found most officers prepared. More rarely, the officers of a regiment in which the lieutenant-colonel was unpopular could deliberately set out to ruin financially their commanding officer, when, by withdrawing from a concordat at the last moment, the lieutenant-colonel was 'allowed to drift into a major-generalship without recovering a farthing back of all that he had laid out'.(32)

In many ways, it was the purchase of promotion, with its undertones of trading, rather than the purchase of the initial commission, which created most controversy. The initial commission was never sold other than at the regulation price, and the amount of money which was invested in the purchase of the initial commission could be equated with the similarly large sums which had to be spent in training the newcomer to any profession. But such an argument had no validity when it came to the question of purchasing promotion, for in no other profession was advancement so blatantly dependent on the possession of money. To many laymen it was only too apparent that the poor man with merit had little chance of advancement. In time of peace it was clear that the number of non-purchase promotions was relatively small. In the years before 1838, for example, it was calculated that, overall, approximately three-quarters of all promotions were filled by purchase (33) (see Table 9). Nor was the position greatly altered at the time of the Crimean War, when it was evident that promotion by non-purchase was unknown in the Guards and most unusual in the cavalry (34) (see Table 10).

Yet in general terms this was not a source of unrest within the military establishment. There were exceptions. In his evidence to the Purchase Commission, Major-General Sir Colin Campbell (Lord Clyde) suggested that many officers put themselves and their families to great inconvenience and embarrassment to avoid being passed over by other purchasers.(35) Sir Duncan MacDougall argued that when officers were passed over because they could not purchase their promotion, this caused the 'most bitter anguish; the very iron going into the souls of the officers'.(36) But it was perhaps significant that many of the officers who were most vehement in the denunciation of the purchase system were those who had reached their own high rank by the operation of the system. Certainly there were always individual cases of hardship. In the 34th Foot, for example, it took Captain Arthur Shawe from 1 April 1836 to 7 February 1851 by non-purchase promotions to reach that rank, whereas a senior captain, John Peel of Cumberland who had purchased his first commission on

TABLE 9 Percentages of purchase and non-purchase promotions: 1834-8

Promotion	Total no. in rank	Promotions no.	Purchase %	Non-purchase %
To Lieutenant-Colonel	254	67	58.8	41.2
To Major	260	149	72.5	27.5
To Captain	1,354	616	80.7	19.3
To Lieutenant	1,952	1,031	76.0	24.0
	3,820	1,863	76.0	24.0

TABLE 10 Promotions 1849-53

Promotion	Total no. in rank	Promotion by purchase			Promotion by non-purchase		
		Cav.	Guards	Line	Cav.	Guards	Line
To Lieutenant-Colonel	181	12	7	63	3	–	28
To Major	257	29	8	124	4	–	60
To Captain	1,413	156	22	566	7	–	193
To Lieutenant	1,966	242	46	998	21	–	652
	3,817	439	83	1,751	35	–	652

22 June 1847, was a Captain by purchase on 25 November
1853. Among the officers of the 30th (The Cambridgeshire)
Foot, Archibald Campbell had been first commissioned as an
Ensign in 1812. He served with the 77th in the Peninsula
from August 1812 to the end of the war, but it was not
until 1837 that he became a Captain and in 1855, when still
a Captain with '32 years service on full pay and 12½ on
half-pay', he was wounded in the assault on the Redan
though almost sixty years of age. Equal problems faced
the officer who was commissioned from the ranks. James
Healey was typical of many more. Healey, who had fought at
the Battle of Toulouse in 1814, was commissioned in the
5th (Princess Charlotte of Wales') Regiment of Dragoon
Guards in 1825. On 25 August 1854 he was promoted to
Captain, six months after William Inglis who had entered
by purchase in 1849. Yet an investigation of the length
of service of officers who had obtained a captaincy as the
last step in a series of non-purchase commissions, showed
that there were considerable variations in the time they
had to wait. The 170 captains in this category had served
on the average for thirteen years and six months before
they received their rank, and while the two longest-serv-
ing officers in the 15th Dragoons and 63rd Foot had served
twenty-three years ten months and twenty-three years six
months respectively, this was balanced by the six years
service of several captains in the 74th and 78th Foot, and
the five years two months of a captain in the 12th
Dragoons.(37)

Many officers, however, were quite prepared to wait for
a non-purchase promotion to the rank of captain, since this
brought to them a sizeable capital sum which they could
realize by the subsequent sale of the commission. Provided
they had served for twenty years they could claim the full
price of their commission, a claim which was never refused
except on account of the misconduct of the officer.(38) A
non-purchase promotion at this rank level could thus be
worth at least £1,100 to a retiring officer in a line regiment. In addition, the officer who had received a series
of non-purchase promotions after an initial 'free' commission, stood to gain a considerable financial reward. Even
if the commission were ultimately sold at the regulation
price only, a not insignificant capital sum could be
realized, as is shown in Table 11.(39)

TABLE 11   Number of officers permitted to sell a
non-purchase commission: 1851-6

Commission	1851-2	1852-3	1853-4	1854-5	1855-6
Lieutenant-Colonel	2	1	-	-	-
Major	-	1	2	-	-
Captain	10	6	9	20	15
Lieutenant	20	29	16	12	11
Ensign	3	5	7	5	-
Total	35	42	34	37	26
Maximum received	£4,500	£4,500	£2,000	£2,000	£2,500
Minimum received	£400	£225	£400	£400	£300

Equally an officer who had purchased a particular
promotion was always happy to receive a subsequent non-
purchase advancement, since this also increased the capital
value of the commission which he could subsequently sell.
And for a young man who attracted the attention of the
military élite, the rewards could not only be financial
ones. Transfers between regiments as non-purchase vacancies arose could bring rapid promotion in the way in which
Garnet Wolseley, an ensign in 1852, was captain four years
later and a lieutenant-colonel by the time he was twenty-
six. This was far faster promotion than by the rule of
seniority, where in the Royal Engineers and Artillery
officers were taking eleven years to reach the rank of
captain and upwards of thirty-seven years to be promoted
to lieutenant-colonel.

Additionally, many of the worst abuses which were associated with the system were gradually checked through
administrative regulations issued by the Horse Guards.
After 1850, for example, certain educational standards had
to be met before an officer was initially commissioned.(40)
The required standard for admission was not particularly
high, although the examination appeared to be a very difficult one since the Horse Guards demanded that in future
all candidates for commissions should pass an examination
in History and Geography, Algebra, Euclid, French, Latin,
Field Fortification, Spelling and Handwriting. The only
comfort which Lord Raglan could offer to a large number

of apprehensive candidates was that a judicious amount of
'cramming' would enable them to master the examination,
but it was noticeable that a number of names were taken off
the list of those seeking commission. Coincidentally it
was expected that, eventually, officers who had at least
provided some evidence of their basic educational ability
would move into the promotion zone. This, it was believed,
would answer the criticism, through excluding from the beginning the least able of the candidates for commissions,
that it was the sot or fool with money who bought his promotion. Equally, to meet the criticism of the inexperienced youthful senior officer, specific periods of service
on full pay had to be completed at a rank level before a
promotion could be purchased. In the early nineteenth
century there had been innumerable examples of these 'boy'
officers. Lord George Beresford, third son of the Marquess
of Waterford, was a lieutenant-colonel at the age of
twenty-three. The Hon. Charles Stewart, second son of the
Earl of Londonderry, purchased the lieutenant-colonelcy of
the 18th Dragoons when twenty-two. In November 1826, the
Hon. George Bingham, later 3rd Earl of Lucan, for whom a
commission had been purchased in the 6th Foot when he was
aged sixteen, purchased over the head of the senior major
of the 17th Lancers to obtain command of the regiment.
At this time Bingham, who was aged twenty-five, had served
with the regiment for eleven months. Previously, he had
served with five different regiments as he moved from the
less fashionable infantry to the more exclusive cavalry.
Yet these rules, irrespective of their effect on the development of military professionalism, only touched lightly
some of the problem areas associated with promotion. One
additional complication, for example, which was peripheral
to the central issue of purchase but still of importance,
was the existence of a dual rank structure. On the one
hand, an officer held a regimental rank, frequently purchased, which was associated with the performance of his
duties in the sub-group of the regiment. These regimental ranks covered the grades of cornet (cavalry) or
ensign (line) through lieutenant, captain and major to
the lieutenant-colonel in command of the regiment. In
contrast an officer could also hold a brevet or army rank
which fixed his position relative to other officers in
the army as a whole, but which was irrelevant in the context of his service within the regiment. These army
ranks began at the field officer level of major, lieutenant-colonel and colonel and finished with the general
officer ranks.
    Brevet rank had been introduced by William III as a
form of reward, but by the time of the Victorian army it

was generally accepted that, with some exceptions, it was
regulated by the rule of seniority. The exceptions
allowed the Commander-in-Chief to award brevet rank to
officers in specific army appointments such as the Staff,
or to bestow brevet rank for meritorious service in the
field, an exception which had become institutionalized
in the Peninsula War. An individual officer could therefore hold two ranks and it was by no means certain that
these would necessarily coincide. In the 76th Foot, for
example, in 1838, the senior major was Joseph Clarke
whose regimental rank had been obtained on 26 June 1833.
The junior major, George Dansey, had a seniority in the
regiment of 26 May 1837, but in the army he was the senior
of the two, for Dansey was a lieutenant-colonel in the
army from 10 January 1837. Similarly, in this regiment
at this time, two of the junior captains, John Chipchase
and Charles Fitzgerald, the most junior of ten, were army
majors. The degree of strain to which situations of this
type could give rise depended very much on the personali-
ties of the officers who were involved. Some accepted
that the situation was anomalous, their interest in their
regiment overcoming any personal reactions. Others, however, were less ready to adopt such an altruistic attitude, particularly in the type of situation in which a
regimental major could find that as an army colonel he
was expected to serve under the direction of a commanding officer whom, in the Army List, he outranked. In
these circumstances considerable strain could develop,
particularly where the brevet army rank had been awarded
for meritorious service in the field, whereas the lieutenant-colonelcy of the regiment had been purchased by a
soldier who had little if any experience of active service. To the officer involved, it was little consolation
to be told by the Horse Guards that others were obliged
to accept the situation. The strain which this dual rank
structure created was, moreover, accentuated during wartime. In 1856, for example, E.G. Hallewell, third senior
captain in the 28th (The North Gloucestershire) Regiment,
served in the Eastern campaign of 1854-5 as Deputy Assis-
tant-Quarter-Master-General to the Light Division in the
army rank of lieutenant-colonel, so outranking his colleagues and both majors in his regiment. It could thus
become very difficult for an officer filling a temporary
appointment on the Staff in his army rank to issue orders
to his regimental colleagues under whom he might be serving again at a later date. Some officers in this
situation delighted in exercising the power which had been
given them, particularly when they legitimately outranked
their former commanding officer or previously senior

colleagues. Others were more diffident, being reluctant
to fulfill the responsibilities thrust upon them by their
appointment and army rank.   In both cases an ancillary
result of the existence of dual structure was the weakened
authority of the Staff Officer and the encouragement of a
potential Staff-Line conflict.

The apparent anomalies created by this dual rank structure became more evident when the recipient of brevet rank
was a Guards officer. From 1776 onwards, officers in the
Brigade of Guards automatically held an army rank higher
than their rank within the regiment. A lieutenant in the
Guards accordingly held the army rank of captain, a captain the rank of lieutenant-colonel and a major the army
rank of colonel. When to this was added the advantage of
brevet rank, the caste privileges of Guards officers frequently irritated many people. As Sidney Herbert when
Secretary of State for War commented, 'I fear a caste
army. I have done something to reduce the privileges of
the Officers of the Guards.'(41) These privileges were
very marked in the promotion and appointments which these
officers received. Sir Nigel Kingscote (1830-1908),
grandson of the 6th Duke of Beaufort, purchased a commission in the Scots Fusilier Guards in 1846 at the age of
sixteen. Four years later, after purchasing promotion,
he was a lieutenant in the regiment with the army rank of
captain. In the Crimean War, as ADC to his kinsman Lord
Raglan, he was a lieutenant-colonel in the army at the
age of twenty-five, although his regimental rank remained
that of lieutenant. This type of accelerated promotion
seemed to negative earlier attempts to avoid the excesses
of purchase, but the more relevant practical problem to
which this privileged position gave rise, was seen at the
time of appointing a successor to Lord Raglan. In April
1854, the 1st Division of British troops comprised a brigade of the Guards and a brigade of Highland regiments.
In the Guards' Brigade, a company of the Coldstreams was
commanded by Sir William Codrington; the Highland Brigade
of three regiments - 42nd, 79th and 93rd Foot - was commanded by Lieutenant-Colonel Sir Colin Campbell with the
local rank of Brigadier-General. Campbell (1792-1863)
was an excellent example of the officer who, lacking
family 'interest', often volunteered for service in the
Indian Army. The illegitimate son of a Glasgow carpenter
and a lady of the Clan Campbell, Sir Colin was first commissioned in 1808 and after serving in the Peninsula War
was a captain in the peacetime army. After purchasing a
lieutenant-colonelcy in 1832 he commanded a brigade in
China during the 1842-6 War, and spent the remainder of
his career in India until going to the Crimea in 1853.

An irascible Scot, Campbell who had served under Moore and
Wellington, had fought all over the world from America to
China. He had been wounded four times and was acknowledged to be both brave and talented. Twice he had .com-
manded a division in India, but at the outbreak of the
Crimean War he was, with forty-four years distinguished
service, still only a lieutenant-colonel. Greatly respected by his men, he was generally recognized to have a
far greater active military experience than the majority
of officers in the Crimea. Both Codrington and Campbell
held the army rank of colonel and both were among the
ninety-five officers promoted to the rank of major-
general in the half-yearly promotions of 20 June 1854.
On the death of Lord Raglan, Codrington, however, was
appointed as Commander-in-Chief of the Eastern Army with
the local rank of general, much to the chagrin of Campbell,
who initially refused to serve under the command of a man
'who at the beginning of the war commanded a company in
the division where I commanded a brigade'.(42)
     In the context of the modern acceptance in the military
of the apparently anomalous effect of promotion by merit
or by selection on the basis of achievement, these nineteenth-century reactions may seem to be unfounded. They
are, however, interesting because they were symptomatic
of a much wider controversial issue. A basic problem in
the Victorian army was to find a satisfactory system of
promotion. There were, it seemed, a number of principles
which could govern the promotion of officers. Each solution to the problem, however, had certain advantages and
disadvantages. First, officers could be promoted solely
on the basis of their seniority. Second, they could be
promoted on merit. Third, they could be promoted according to some combination of these two concepts. Finally,
the existing system could be retained. The first solution
had the advantage that it extended to all ranks, both in
the army and in the regiment, the principle which was
already applicable to promotions in general officer grades
and to many brevet rank appointments. It was a known and
tried system which possessed the additional advantage of
being certain and free from bias, while, if the system
were adopted, the establishment could still reward selected officers by giving them additional seniority. It was
questionable, however, whether these advantages outweighed
the major disadvantages. There was no reason to suppose
that the adoption of this principle would encourage the
development of professionalism, and it could be argued
that a system which guaranteed an officer's promotion,
would provide little incentive for individual initiative.
Certainly experience in the operation of the seniority

rule within the corps of Royal Artillery and Royal Engineers, and in the general officer grades of the army, suggested that the positive disadvantage of promotion by seniority was the way in which it produced a group of over-age officers many of whom were no longer able because of their age to carry out the duties expected of them. It was only necessary to refer to the difficulties faced by the military establishment in finding general officers to serve in the Crimea, to appreciate the very real problems which the adoption of the principle of promotion by seniority could give rise.

The concept of promotion by merit was equally associated with specific advantages and disadvantages. To the layman it had the attraction of establishing within the army a principle which had already been put forward for the Home and Indian Civil Service and it satisfied a Utilitarian ideology which was the basis of many Victorian reform movements. The advantages seemed to speak for themselves - the development of professionalism, the selection of officers with ability, and the ending of privilege - but both in the army and in society there were doubters who preferred to adopt a more cautious approach. Their arguments were often unpopular with those critics of patronage who approached reform with an almost religious fervour, since they appeared to be an attack on a spirit of Puritanism which laid great stress on the moral value of success in a chosen occupation.(43) Apparently these cautious men were unwilling to accept without question the argument that 'well-educated poor men are notoriously those who throw themselves into their work with the greatest energy and perseverance'.(44) They were, however, in the context of the period, practical arguments which removed the debate from an area of theoretical abstractions to a more mundane pragmatic level. De Fonblanque, although an advocate of reform, made this very clear when he drew attention to the very real difficulties associated with a system of promotion based on merit:(45)

> Constituted as our Government is, the claims of political supporters and adherents are so strong as to influence most powerfully the patronage of the executive, and although pre-eminent services might as a rule be recognised and rewarded, the claims of party would under ordinary circumstances assert themselves above all others, and in many cases supercede the superior claims of service. This liability to the abuse of patronage, which arises less from the faults of our public men than from the nature of our institutions, operates as powerfully upon the army as upon other branches of the public service.

A similar reaction also occurred among those army
officers who were 'reformers', a change of attitude which
frequently resulted from their experiences in the Crimean
War when they were able to make direct comparisons between
the effects of promotion by merit in the French Army and
promotion by purchase in the British Army. Thus Major-
General Sir John Burgoyne, who from 1834 onwards had urged
the abolition of the purchase system, revised his opinion
after his experiences in the Crimea led him to conclude
that the French system of promotion by merit showed how
officers could become involved in politics. Promotion by
merit, he argued, resulted in dissatisfied officers, and,
in addition, since 'merit' could so often be equated with
political influence, this contributed to military weakness.(46) In retrospect, these fears seem to be exaggerated and an example of the way in which the conservative
officer sought to rationalize his reaction to proposed
changes in the system by describing selection as favouri-
tism and by making the most of its sinister possibilities.
Yet in the historical context of these proposals, the
army's fears were not entirely irrational, even if the
case was over-stated by military preservationists. General
military experience of the effects of political influence
in the selection of officers during the eighteenth century, and specific experience of the way in which selection for élite appointments in the nineteenth century had
been affected by considerations of political influence and
family interest, validated their reaction to the association they made between selection by merit and political
intrigue. But their reaction was also founded on more
pragmatic reasons. An initial problem was the definition
of 'merit' within a military context. Was this to be
equated with concepts such as 'bravery', 'honour', or
'heroism'? This was undoubtedly how most laymen interpreted military merit, admiring men like Sir Colin Campbell
who when ordered to withdraw at the Alma during the Crimean War, replied brusquely, 'No Sir! British troops never
do that nor ever shall while I can prevent it.'(47) As
'The Times' commented in a leading article,(48)

There is a cant of democracy as well as of aristocracy,
and merit, the only true criterion is as little satisfied by promoting a man because he is poor as because
he is rich . . . what we desire is first, that a place
in the Army shall be open to everyone whose bravery and
intelligence qualify him to fill it.

Was merit to be interpreted as 'leadership' or 'efficiency', qualities which were themselves almost incapable
of further definition, or was it to be regarded as something akin to the 'academic ability' which was associated

with the promotion of individuals in the civil bureaucracy? None of these were qualities which necessarily
led to success against an enemy, perhaps the only valid
evaluation of 'merit', and there were innumerable examples
of instances where officers possessing these characteristics of bravery, leadership, efficiency and academic
ability had failed miserably in a war situation. One of
the outstanding and most tragic examples of these failures
subsequently occurred during the First South African War
of 1881-2. The commander of the British forces sent from
Natal to relieve the garrisons beleaguered by the Boers
in the Transvaal was Major-General George Pomeroy-Colley
who, at the age of forty-six, was widely regarded as the
most promising general in the post-1870 army. Not only
had he passed out top of his class at Sandhurst when aged
sixteen, but he had also passed out first at the Staff
College with a record total of marks, although he had preferred to avoid lectures and work on his own, sitting the
final examination after one year instead of the normal
two. This academic brilliance was supplemented by an
outstanding administrative ability which he had displayed
through his contribution to the Cardwell reforms and in
the Ashanti War of 1873-4 where he had taken charge of
communications. By 1880, when he was Governor of Natal,
Colley seemed to be the epitome of the successful professional general — well educated, efficient, highly intelligent, commanding the respect of his subordinates and
incredibly self-assured. The first occasion, however,
when he was given a chance to exercise independent command
in the field ended in disaster. After suffering two heavy
defeats in the minor actions of Laing's Nek and at the
River Ingogo, when a third of the British force became
casualties, Colley took seven infantry companies and a
naval detachment of 554 men to the top of the apparently
impregnable Majuba mountain (6,500 feet), believing that
this move would force the Boers to quit their positions
which he now dominated and thus open the route into the
Transvaal. Between 1.00 p.m. and 2.30 p.m. the following
day a storming party of Boers took the summit, completely
overpowering the British troops. Among the 93 killed and
133 British wounded was Major-General Colley himself,
beaten by the instinctive grasp of tactics and superb
marksmanship shown by the Boers.(49)
    A second reason for the caution with which the officer
corps accepted the claims put forward to justify the introduction of promotion by merit was derived from their own
experiences in preparing confidential reports. In accepting the principles of staff evaluation, the Victorian
military establishment was more than a hundred years

ahead of most British industries and public administration. This early reporting system lacked the sophistica-
tion of the modern scientifically designed system currently
used in the British military, but it was considerably in
advance of other Victorian situations. (50) The principles
which were laid down in Chapters 189 to 192 of 'Queen's
Regulations' placed the major responsibility for the
making of these reports on the general officers who carried out the half-yearly inspection of units. These
officers were urged to bring to the attention of the
Commander-in-Chief those field officers and captains who
must be superior to others, and in this context, regimental commanders were under a duty 'to bring especially
to the notice of the inspecting general without favour or
partiality any officer who may be distinguished for
attention to and proficiency in his duties'.(51) Conversely, the inspecting general and commanding officers
were 'to particularise those officers who from incapacity
or habitual inattention, are deficient in a knowledge
of their duties, or who show an indisposition to afford
to the commanding officer that support which he has a
right to expect from them'.(52) The problems which the
reporting officers could encounter were appreciated and
they were urged to give their opinions 'fully, fearlessly
and conscientiously on every matter brought to their
notice, or coming under their actual observation, whether
it be one requiring praise or censure'.(53)
    While this suggested that 'merit' could be defined
satisfactorily as 'attention to and proficiency in
duties', the difficulties which arose in implementing the
regulations suggested that there would be comparable
problems if promotion were to be based on merit. It was
evident, for example, that there was a tendency toward
negative evaluation, partly because it was safer to
criticize than praise, and partly because the power of the
veto was, as the Commander-in-Chief told the Purchase
Commission in 1856, 'much more easy to exercise than the
power of selection'.(54) It was also clear that a feeling
that evaluation was an attack on the corporate spirit of
the regiment generated considerable hostility towards the
implementation of the reporting system. This had been
cogently summed up by Lieutenant-General Blakeney:(55)

    The confidential report has turned the Army whose
    constitution is based on the most scrupulous adherence
    to the highest and nicest principles of honour, into a
    graduated corps of spies from the ensign up to the
    general.
But an argument of this type, although it hinted at the
dysfunctional consequences which could arise from the

introduction of merit evaluation, was irrelevant in the
eyes of those reformers who naively believed that 'fitness
for command of a regiment is such a well marked qualification' and who emphasized the importance of factors such as
'energy' and 'perseverance'.(56)

In this controversy, there were weighty arguments in
favour of the retention of the status quo. Not only was
the system self-perpetuating and a contribution to stability in an era of change, but it was encouraged by the
Crown. It ensured that officers would be linked to that
very important basis of property which also characterized
eligibility for public office in fields as varied as the
Militia, the House of Commons and the Magistracy. It
meant that officers belonged to the 'right' class, and by
implication, the right class could only be obtained if
the barriers to their recruitment were of minimal effect,
a condition which could most easily be achieved by the
retention of some system of patronage whereby men most
fitted by birth, breeding and education were those recruited into the officer corps.

Underlying this connection between 'officer' and
'tentleman', was a belief, derived from seventeenth-
century convictions, that a professional standing army was
a menace to English liberty. There was a very real fear
that if the independent gentleman was not encouraged
through the purchase system to take up a military career,
he would be replaced by men who had little interest in the
well-being of the country, and who would be predisposed to
use the army as a weapon in internal politics. The real
value of the system, as it appeared to the traditionalist,
had been put very clearly by Wellington in 1830:(57)

> It brings into the service men of fortune and education
> — men who have some connection with the interests and
> fortune of the country besides the commissions which
> they hold from his Majesty. It is this circumstance
> which exempts the British Army from the character of
> being a 'mercenary' army, and has rendered its employ-
> ment for nearly a century and a half, not only incon-
> sistent with the constitutional privileges of the
> country, but safe and beneficial.

This belief that the system contributed toward the subordination of the military to the civil power was repeatedly put forward as a reason for its continuance. In
1856, for example, Lord Palmerston argued very forcibly
in the House of Commons in favour of the purchase system:
(58)

> It was very desirable to connect the higher classes of
> society with the army . . . if the connection between
> the army and the higher classes of society were

dissolved, then the army would present a dangerous
and unconstitutional appearance. It was only when the
army was unconnected with those whose property gave
them an interest in the country, and was commanded by
the unprincipled military adventurers, that it ever
became formidable to the liberties of the nation.

This was a sentiment which was consistently repeated,
but two other, more pragmatic reasons made it an attractive system in the eyes of the Crown. The purchase system
acted as a guarantee of good behaviour, since an officer
who was cashiered from the army forfeited the amount of
the price he had paid for his commission. In this respect
the initial price which was put down acted as a fidelity
bond and, seen against the Royal Warrant of 1821, the
amount to lose could be very substantial. Critics of the
system argued that it was very unusual for an officer to
forfeit the value of his commission through misconduct,
since an individual who was to be dismissed from the army
was 'invited to retire', a procedure which allowed him to
sell out. Nevertheless there was always a latent sanction
available to the military establishment and there were
instances when it was used against an individual officer.
More significantly, the system enabled the Crown to avoid
its duty and responsibility to provide a reasonable reward, and to grant an adequate pension on the retirement
of the officer. In the former case, there was no great
increase in the pay of officers from the seventeenth to
the nineteenth century. In the latter instance, it was
accepted that the money realized by the sale of a commission was in lieu of a paid pension.(59)

The pay was always looked upon as an 'honorarium' and
not 'merces'. It was, as Fortescue has suggested, in the
nature of a retaining fee against the day of prize
money; 'less a wage for service done than a fund to enable the service to be performed; rather the tools of a
trade than the profits of a trade'.(60) The rates laid
down in 1797 remained unchanged during most of the nineteenth century, but surprisingly, while this lower rate
was a frequent source of complaint, it was never a major
controversial issue.(61) To a foreign observer of the
Victorian military establishment, it was surprising that
these complaints were so restrained in their nature, for
it appeared to him that the low rate of reward was a
potential source of disaffection:(62)

While the progress of the Arts spread through almost
every class of society, the enjoyments of comforts
hitherto unknown, military officers were obliged to
impose on themselves increasing sacrifices, which were
rendered the more painful by the contrast around them.

Their subordination, however, was such that no tumult,
nor revolt, scarcely even a complaint, betrayed the
discontent of the army.

There was certainly no class of a comparable social
status, with the possible exceptions of the naval officer
and the country clergyman, who had to work for so little.
The average income of other professional groups was far
higher. As Fox Maule (subsequently 2nd Lord Panmure and
11th Earl of Dalhousie) pointed out in the House of
Commons in 1850, it was his experience that the officers
of the British Army were the worst-paid and the hardest-
working of public servants. He asked the House to consider the position:(63)

> Look to the pay of officer of a regiment, and in the
> first place to the pay of a lieutenant-colonel. I will
> treat it in a mercantile way, so that it may be per-
> fectly plain to the understandings of mercantile men.
> The lieutenant-colonel to arrive at that rank in the
> army, paid £4,500 for his commission; and his pay for
> commanding the regiment was £365. If you deduct from
> the price of his commission, the interest at 5%, which
> was but a fair deduction amounting to £220, and regi-
> mental expenses at £20, which he had no alternative but
> to incur, and deduct the income tax at £11 on his pay,
> it would in all amount to £258 leaving the sum of £107
> as the pay of a lieutenant-colonel for the duty he
> undertakes.

Since, however, the regulation price of a commission was
rarely paid, because officers were forced to purchase at
a price far in excess of the commission's nominal value,
the actual income of an officer, calculated in accordance
with the formula adopted by Fox Maule was far less than he
had suggested. From the figures which are quoted in
Table 12,(64) it can be seen that there was more than a
little truth in Wellington's dictum that 'three-fourths
of the whole number receive but little for their service
besides the honour of serving the King'.(65)

Evidence to the 1838 Royal Commission on Naval and Military Promotion, moreover, showed that even if the interest
on the capital cost of the commission was ignored in calculation, many officers could not meet their expenses from
their military pay:(66)

> From the pay of an ensign which was 5/3 a day, his din-
> ner will come to 2/6, and his breakfast will come to
> 1/0, and then with the residue he has to clothe him-
> self. . . the expenses that an officer is exposed to,
> are certainly more than his pay will meet.

TABLE 12   Real income derived from a purchased commission

Rank	Pay and allowances	Interest and expenses	Differences
Lieutenant-Colonel	£365 0 0	£380 12 11	−£15 12 11
Major	£292 0 0	£249 10 4	+£42 9 8
Captain	£211 7 11	£142 14 11	+£68 13 0
Lieutenant	£118 12 6	£53 19 5	+£64 13 1
Ensign	£95 16 3	£27 15 0	+£68 1 3
Average	£216 11 4	£170 18 6	+£45 12 9

Writing in 1867, when the pay of an ensign was still only
5/3 a day and when some former voluntary payments, such
as twelve days pay a year for the band, had become formalized, Trevelyan suggested that the parents of young
officers, besides purchasing their commissions and fur-
nishing their outfits, had to make them an annual allowance for subsistence.(67)   Additionally, if the officer
was serving in a fashionable regiment where a colonel
insisted on elaborate changes of uniform, the cost of an
officer's outfit could equal four years' pay for a young
ensign, or as much as a lieutenant-colonel's pay for a
year.   These costs are brought out clearly by Thackeray
in 'The Book of Snobs', where he prints the outfitting
bill for Lieutenant Wellesley Ponto of a mythical 120th
Queen's Own Pyebald Hussars.   It is not, however, grossly
exaggerated (see page 88).
    The accepted justification for this low rate of reward
was derived from the premise of Adam Smith that 'Honour
makes a great part of the reward of the honourable professions'.(69)   Military reward, however, was so much
lower than in other comparable professions, that it
severely limited the field of applicants.   A comparison
between the salaries of army officers and civil servants
in the War Office of equivalent rank, shows that the re-
muneration of the former was less than half that of the
latter.   A lieutenant-colonel who received a gross pay of
£365 a year can be compared with his civilian counterpart,
a clerk 1st class, who was paid on a salary scale which
ranged from £670-800 a year.   Similarly, while an army
captain drew £211 per annum, his equivalent, a clerk 2nd
class, enjoyed a salary on a scale of £315-500.   There
was thus very little financial inducement for an officer

Out fitting bill for an officer of the Hussars

	£	s	d		£	s	d
Dress Jacket, richly laced with Gold	35	0	0	Brought Forward	207	3	0
Ditto Pelisse, ditto, and trimmed with Sable	60	0	0	Gold Barrelled Sash	11	18	0
				Sword	11	11	0
Undress Jacket, trimmed with gold	15	15	0	Ditto Belt & Sabretache	16	16	0
				Pouch and Belt	15	15	0
Ditto Pelisse	30	0	0	Sword Knot	1	4	0
Dress Pantaloons	12	0	0	Cloak	13	13	0
Ditto Overalls, gold lace on sides	6	6	0	Valise	3	13	6
				Regulation Saddle	7	17	6
Undress ditto ditto	5	5	0	Ditto Bridle, complete	10	10	0
Blue Braided Frock	14	14	0	Dress Housing, complete	30	0	0
Forage Cap	3	3	0	A Pair of Pistols	10	10	0
Dress Cap, gold links plume & chain	25	0	0	A Black Sheepskin, edged	6	18	0
	£207	3	0		£347	9	0

to join the Victorian army and while a complex system of allowances complemented the basic rate of pay,(70) there was little reason for the Crown to fear that mercenary officers, 'prone to disaffection and subordination', would be recruited into the military for the extrinsic rewards which they would receive.

In a similar way, the existence of the purchase system was encouraged by the Crown because it avoided accepting any responsibility toward paying a pension to the super-annuated officer. In contrast with Weber's dictum that a characteristic of a bureaucratic administrative staff was the principle that they should be 'remunerated by fixed salaries in money, for the most part with a right to pension',(71) the Victorian military establishment de-pended on a neo-feudal system of grace and favour. Ini-tially a total sum of £21,000 a year, derived from the commutation in 1806 of certain garrison appointments which carried a daily or annual rate of pay, but involved no performance of duties, had been available for distribution. This, in the 1830s, was shared among 131 officers, although under a royal warrant of 1 October 1840, 20 lieu- tenant-colonels, 20 majors and 115 captains could be re- tired on full pay. Even after the Crimean War the retired list of officers of the rank of major and above who were paid a pension for their former military service was rela- tively small. The Army List in 1856, for example, totalled 652 field and general officers (2 lieutenant-generals, 12 major-generals, 122 colonels, 288 lieutenant-colonels and 228 majors). Moreover, this pension unless it had been awarded for meritorious service in the field was not paid as 'a right' but was dependent on the applicant satisfying a complex system of qualifying criteria.

For the remainder of the officer corps, the basic al- ternatives were to sell their commission and recoup the invested capital amount, or to go onto half pay. Either alternative was characterized by a large number of his- torical anomalies, many of which were derived from dubious precedent. If an officer were killed, for example, his commission could not be sold by his heirs, and no payment was made to his widow, although the Horse Guards could, at the discretion of the Commander-in-Chief, make an ex gratia peyment. Evidence given to the Purchase Commission showed that a considerable number of small sums of money were paid on the authority of the Commander-in-Chief to rela- tives, living in distressed circumstances, of deceased officers. After 1856, the commission of an officer killed on active service could be sold on behalf of the widow of the deceased officer, but complaints were made to the Purchase Commission that the widow had to show her 'need' for the money and that it was not paid over as a right. To avoid the possible loss of this capital sum, many officers insured the value of their commission with one of the insurance companies which specialized in providing cover, although the premium, at an annual rate of £2 10 0 to £5 0 0 per £100 assured, was very heavy in relation to the officers' income. Whole life cover in 1856 for an officer who had purchased a lieutenant-colonelcy for £14,000 could amount to a premium of £700 per annum against pay and allowance of £365. For a young captain in a line regiment whose commission cost £2,400, the respective figures were premium £60, and income £211 a year. Addi- tionally, the officer promoted to the rank of major- general could not, as we have seen, sell his commission, and if he could come to no arrangement with his regimental colleagues he was forced to rely on some other source of income, possibly from a State office, as an alternative to a pension.

Many officers chose to go on half pay. While the latter was in many ways comparable with the modern concept of a pension related to occupational earnings, it could not be enjoyed by an officer who sold his commission. Essentially, it was a 'retaining-fee', which was only paid to the individual so long as he held a commission and was, in theory if not in practice, available for further service. From time to time the half-pay list was shortened by bringing officers back on to full pay and then 'forcing' them to retire by selling their commissions, since they were not allowed to go back onto half pay. Half pay was disliked by the Treasury because of its high cost — £250,000 a year — and it created further anomalies which reflected on any evaluation of military professionalism. The officer on half pay continued to receive promotion, and it was possible for an officer to be promoted to the rank of major-general with as little as seven or eight years' service on full pay. In the promotions which took place between 1830 and 1838, 123 officers were promoted to the rank of major-general. Of these, 2 had less than ten years' active service; 23 had less than twenty years' active service, and a further 21 had less than twenty-five years' active service. In the August 1855 promotions, sixteen of the seventeen promoted officers were veterans of the Napoleonic War who had gone onto half pay before 1835. As such, they enjoyed the half pay of their last regimental commission so that James M'Haffie, first commissioned in August 1797, had received half pay as a Captain in the 60th Foot from November 1818 onwards. He was not, however, atypical, for Ambrose Lane, a Peninsular veteran who had first been commissioned in 1783, had gone on to half pay in January 1821 and had been promoted to the rank of major-general in May 1855, seventy-two years after he had been first commissioned and after thirty-four years of inactive service. Some officers who were promoted in 1855 were on the half pay of regiments which no longer existed. Rodolphe de May of Watteville's Regiment, Sir George Julius Hartmann who had been an artillery major in the German Legion until 1816, James Horton of Meuron's Regiment, and Colin Pringle of the 6th Battalion, German Legion, were officers of foreign regiments which had been raised for service in the 1800s. This promotion debased the value of the rank and it suggested that promotion was in no way related to either merit or availability for service, but was simply a reward for loyalty, in this case inactive loyalty, to the military establishment.

In making a decision whether to go onto half pay or not, the officer was thus faced with a choice. He could in theory either sell out, and in recouping his investment

in the system coincidentally sever his connections with
the military, or he could retain his commission and enjoy
half pay at the risk of losing the capital value of the
commission on his death or on promotion to general rank.
In any event, no adequate provision was made for the payment of a pension, so that the retention of the purchase
system relieved the civil power of its obligation to
provide for officers on the basis of the service which
they had rendered to the State.

For a considerable part of its lifetime, the Victorian
army was thus involved in debate about the extent to which
the retention of the purchase system retarded the development of professionalism. In this controversial area,
'reformers' and 'preservationists' could equally put forward arguments to justify the adoption of their point of
view using examples of individual officers in the army
whose experiences in terms of success or failure seemed
to confirm the validity of their solutions to the problem
under debate. It was an area of controversy in which,
from time to time, arguments could become bitter, for to
the military, the attack which was made on their system
appeared to be an attempt to destroy their corporate
entity. This was a logical reaction since the officer
corps sought to perpetuate its élite self-conception,
striving to ensure that the distinctive quality of military life was neither subjected to critical evaluation
nor eroded through the admission into the group of outsiders. Equally, the point of view put forward by the
reformers was an understandable attitude.

It was essentially a logical reaction to the continuance of an eighteenth-century anomaly in nineteenth-
century society. The cry for reform echoed the demands
for reform in the Civil Service, and for changes in other
spheres of public activity such as education and local
administration. It reflected the demands of a growing
professional middle class for the opportunity and the
right to become members of the military establishment,
particularly since the arguments which had previously
been put forward in support of the maintenance of the
system made less and less sense as the century progressed.
Now, as Trevelyan argued, it was necessary to increase the
power of the rising middle class:(72)

> The object to be aimed at is to make the army a true
> representative of the nation. It should be neither
> more aristocratic nor more democratic than the rest
> of English society. The upper, the middle and the
> lower classes cordially co-operate in every other pub-
> lic and private undertaking, and why should the army
> be the solitary exception? Since the beginning of the

> century, the middle class has enormously developed,
> the present flourishing state of the country is
> chiefly owing to their exertions.

This was a very attractive argument, not least because in
a period when merchants and industrialists were not yet
automatically accepted into society, a military career was
one way in which the status aspirations of a professional
class could be satisfied. It also mirrored other attempts
made by the middle class either to achieve economic power
through 'the gradual enlargement of the interests of commerce and the only means by which human liberties can ever
be built up upon a secure foundation',(73) or to participate more fully in the exercise of political authority,
either directly through an enlarged franchise, or indirectly through membership of the Civil Service.

The argument in favour of a widened base of recruitment
could be evidenced, moreover, by historical example:(74)

> In former days, the middle class was trained to the
> use of arms with a view to national defence, and they
> showed on many memorable occasions, what they were
> capable of for the honour and safety of England . . .
> under Cromwell they held all England, Ireland and
> Scotland in subjection. . . . In recent times, the
> middle class has received a remarkable extension. . . .
> To allow no place to this portion of our population in
> the army is like fighting with one hand tied.

To allay the fears of those members of English society
who still feared the threat of a mercenary army which
might attempt to intervene in the conduct of domestic
politics, it was argued that the abolition of the purchase
system would increase rather than diminish state control
over the military establishment. The drawback of the
existing system was that it encouraged commanding officers
to look upon the regiment as their personal property. With
the abolition of the system, it would be possible to 'give
the state that unrestricted power over the officers of the
Army which it is desirable that it should possess'.(75)
But the impact of this argument on the decisions finally
made by the political élite, like those advanced by
De Fonblanque, Higgins and Trevelyan in favour of greater
middle class participation in the military or in support
of improved military professionalism, must be treated with
a certain amount of caution. There was a demand for
change. It was most marked at the end of the Crimean War
during the late 1850s; it was apparent again ten years
later. To a certain extent it mirrored the demands of
contemporary society, although a large body of the military
and the public were resolutely opposed to it, from Queen
Victoria who feared that promotion by selection rather

than purchase would be very difficult,(76) to military
conservatives who rejected any amendment of the status
quo. But the crucial question was whether the purchase
system could be retained at a time when there was a pres-
sing need for large scale army reform. This was very
clearly brought out in 1868 when the second 'reforming'
administration of the nineteenth century came into office
under Gladstone, with Cardwell at the War Office. The
latter's task, conditioned by the military effects of the
American Civil War and subsequently affected by the brilliant Prussian success of 1870, was to create a military
establishment which could meet Britain's dual strategic
roles as both a European and an imperialist power. To
do this, he needed to 'localize' the military in territorial areas, where line regiments of the regular army
could be linked with militia regiments recruited locally.
Accordingly, he was obliged to abolish the purchase system, for this barred regimental re-organization and created
a rigid regimental structure. As Cardwell argued in the
House of Commons,(77)

> Do you wish to increase the number of double battalions
> with a view to the Indian branch of the army, and to
> short terms of service — a point of the greatest
> possible importance? If you do so wish, you will be
> met immediately by difficulties arising from the
> purchase question. . . . Do you wish to unite closely
> the militia and regular forces? If you do, one of the
> first things you will have to do will be to give sub-
> altern officers of the militia commissions in the line
> without purchase, and how can this be done if there
> remains any condition in reference to the purchase
> system?

So, when the purchase system was finally ended by royal
warrant, the change was motivated by needs of the organization rather than by the arguments about professionalism
which were advanced by the reformers. Even so, the long
and often bitter controversy over the existence of the
system still affected external evaluation of the effects
of its abolition. It excited more notice than its importance in Cardwell's reforms merited, a reaction which subsequent commentators have elaborated upon, until the other,
and more important, reforms which Cardwell instituted
have tended to be pushed into the background. Thus
Fortescue, for example, in his monumental history of the
British Army, chose to mark the end of the old army with
the passing of long enlistments and purchase of commissions, rather than with reference to either Cardwell's
later reforms or to those of his successors.(78) Additionally, contemporaries, whether they supported or

opposed the system, concluded that the abolition of the
system would bring considerable changes in the composition of the officers corps and in the degree of military
professionalism.(79) It seemed that the criticism made
in the 1856 Royal Commission that purchase was 'vicious
in principle, repugnant to the public sentiment of the
present day, equally inconsistent with the honour of the
military profession and the policy of the British Empire,
and irreconcilable with justice', had been vindicated.

But several questions remained unanswered. First, it
could be questioned whether the abolition of the purchase
system would encourage, ipso facto, a development of military professionalism. This conclusion, as we have seen,
had been assured by the reformers to be a logical result
of the removal of patronage from the military establishment, but they had chosen to ignore in their discussions
the specific complexities of professionalism in this area.
This was a problem which was closely associated with the
status of the military as an organization, and in view of
the extent to which profession and organization had fused
in the Victorian military, any consideration of changes
in the former raised the additional question of the effect
of the abolition of the purchase system on the military as
an organization. A second question which arose was the
effect of the amended system on the political attitudes
of the military. Potential changes in the relationship
of the military to the civil power had been largely ignored by the opponents of the purchase system, although
supporters of patronage had drawn attention to the dangers
which might arise, as an argument in favour of retaining
the status quo. Yet this was a question of the utmost
importance, for if the abolition of the purchase system
destroyed the identification of the military with the
landed interest, with what other section of the society
was the military to identify? Was this to be a rising
middle class, a class which was beginning to change the
rules which had hitherto governed the involvement of the
military in the political system? Or would the inability
of the military to develop a much more sophisticated understanding and approach in its relationship with this growing political force, drive the army into a situation where
it became a self-contained, relatively isolated social
system whose believed interests were different from, if
not opposed to, those of the civil power?

Yet all these questions stemmed from the initial presumption that the abolition of patronage, through the
rejection of the purchase system, would greatly change the
composition of the traditional field of recruitment to the
officer corps. Such a presumption was indeed made by many

people. A contemporary commentator writing in the 'Fortnightly Review' in 1886, thought there was no question of
recruitment following a traditional pattern. He concluded
that (80)

> The abolition of promotion by purchase, the institution
> of tests of efficiency; the shortening of the soldier's
> service in the ranks . . . all tend towards making the
> army less a class and more a popular institution stand-
> ing on the broad basis of democratic goodwill.

But this was an over-optimistic view of anticipated
changes which was not endorsed by subsequent factual evidence. The statistics by Razzell and Otley, to which
previous reference has been made, show that abolition did
not open commissions to merit alone. They also emphasize
the extent to which a number of regiments became more
rather than less exclusive. Equally, commissioning from
the ranks remained unlikely. Sir William Robertson,
writing of his life as a Troop Sergeant Major in the 16th
Lancers in 1885, commented that apart from riding masters
and quarter masters, it was very seldom that anyone was
promoted from the ranks — not more than four or five a
year on average. Indeed, he argued that during this
period 'the ranker was not as welcome to the officers of
a regiment as before',(81) a point of view which was
partially evidenced by the way in which the post of
adjutant in a regiment, hitherto used to accommodate a
ranker, was by 1885 being filled by young officers.

After 1871, despite the abolition of purchase, the army
continued to be an exclusive élite group so that the
changes which many reformers had forecasted did not in
fact take place. What had been overlooked in all the
arguments which had been advanced, was that entry into the
officer corps had never depended solely on the ability of
the individual to buy a commission. In common with other
groups in society, the military had developed a group culture, which in this particular instance was further differentiated into the sub-cultures of specific regiments.
To maintain the integrity of that culture, other criteria
of selection had always been employed. Thus in élite formations such as the Brigade of Guards, a culture derived
from a common extra-military life-style and from behavioural interaction had always been maintained through the exclusion of outsiders. What happened after 1871 was that
two of the criteria which had been used to supplement the
sieving effect of the purchase system became increasingly
important, until they completely replaced the restrictive
elements of the abolished system. The potential area of
recruitment might have been enlarged, but now the two
barriers of 'previous educational experience' and 'financ-

ial standing' were constantly used to limit the broaden-
ing of the socio-economic base of recruitment. The former
supplemented other ascriptive criteria, so that attendance
at a particular type of school, usually the public school
and particularly the so-called 'Clarendon' schools, became
a prerequisite of admission to many regiments. The social
network of kinship or membership of the landed interest
was thus increasingly extended in the last quarter of the
nineteenth century to include this peculiarly British 'old
school tie' relationship. As Captain Cairnes commented
in 1900 in giving advice to the military aspirant: 'Being
a young man of discretion, and probably having friends or
relations, certainly old schoolfellows in the regiment of
his choice, he will put himself in communication with his
corps.'(82) This educational experience, however, cannot
be interpreted as a potential indicator of increased military professionalism, for it was largely irrelevant in
terms of academic attainment and it was stressed that
'prominence in athletic exercises is the surest road to
pre-eminence',(83) since character rather than intellect,
and brawn rather than brain remained the criteria of
success.
   The importance of this shared educational experience to
the military was therefore no way related to the possible
development of a meritocracy, and its main significance as
a criterion of eligibility for commissioning was that it
was thought to ensure that the aspiring officer had been
educated to an acceptable standard of social fitness
through the scholastic socialization process. The latter
in this period thus usually supplemented or, more rarely,
replaced the family socialization process, so that both
processes ensured a group cohesiveness which, it was
argued, was a fundamental necessity in a military force.
Such a formulated cohesiveness in which, for example, 'The
background of officers of the 2nd Scottish Rifles was very
similar. They came from about ten public schools — Eton,
Harrow, Winchester and Wellington predominating',(84)
excluded outsiders. It was the basis of a curiously inter-
related public school and military attitude towards such
concepts as 'honour' and 'esprit-de-corps'. So in those
rare cases, 'Where a number of very undesirable young men
find their way into the army, having as a rule nothing to
recommend them but the riches which their parents have
acquired in trade', this group cohesiveness could be relied
upon as a sanctions mechanism. 'We can safely leave these
young men to the tender mercies of their brother officers.'
(85) This refusal or unwillingness to conform to the customs of the regiment invited at best social ostracism, or
at worst a degree of 'hazing' and bullying designed to

drive the nonconformist from the regiment. In this, the
unofficial 'subaltern's court-martial' was frequently used
to punish the apparent deviant, so that any formal approach
to the problem of maintaining group standards was reinforced by the informal code which was applied in practice.

In addition, the barriers erected by the military
against open recruitment were strengthened by the way in
which the officer corps took into account, as a criteria
of eligibility for commissioning, the financial standing
of entrants. The extent, therefore, to which educational
criteria, in common with the defunct purchase system reduced the size of the potential area of recruitment into
the officer corps, was supplemented by the way in which
the expenses of the mess and the more general costs of
military life served to deter many aspiring officers. It
was generally accepted that at no time and in no place
could the British officer in these years depend on his pay
as his only means of subsistence. Many regiments indeed
laid down a recommended scale of private incomes which
young officers were to possess. In the more exclusive
regiments of the cavalry, £600-700 a year was necessary.
(86) Line regiments reduced this to an annual income of
£100, but such a rate was the mandatory minimum and the
recommended rate was often more than this. As Baynes
comments,(87)

The average officer in the 2nd Scottish Rifles had a
private income of about £250 a year. The Regiment
itself only insisted upon £100 a year, though £200
was recommended. This needs to be seen in perspective.
The Coldstream Guards at this time considered £400
a year to be the minimum for an ensign on joining.

It was however, possible for an officer to manage on less
than this, but as Peter Laslett points out, while it is
difficult to decide how much money was needed to maintain
a 'solid middle-class' standard of living at this time,
possibly 'seven hundred or a thousand golden sovereigns
a year' was required.(88) Since no more than about
280,000 households out of a total of 7,000,000 enjoyed
such an income, the area of potential military recruitment
was perforce very limited. At the same time, regimental
insistence on officers possessing a large private income
could be rationalized on the grounds that without such an
income, it would be impossible for officers to enjoy the
normal standard of living of this privileged minority.
Thus, all officers were expected to live up to the standard of their regimental mess and to bear a due share of
expenses incurred by the entertainment of mess guests,
balls, race-meetings, and so forth, even if this created
problems for the individual. 'Officers', warned Cairnes,

'have lived in the 10th (Hussars) with an allowance of
only £500 a year but they have rarely lasted long.'(89)
The case of two young officers of the 4th Hussars who were
bullied and hazed because they were unable to live in the
manner expected from them by their fellow officers was
brought before Parliament in 1896.(90)   The ostracism they
suffered was akin to that which excited 'The Times' in the
1850s when, among others, Cornet Baumgarten of the 6th
Dragoons — 'a Bachelor of Arts of Oxford, a good oar, a
first-class cricketer, a bold rider and a pleasant companion' was so bullied by his fellow officers that he was
forced to quit the regiment.

When to these amounts was added the cost of outfitting
the newly-commissioned officer at about £200 for the infantry and £600 to £1,000 for the cavalry,(91) it was
evident that the army was still an occupation open only in
general terms to the moneyed classes, 'the sons of merchants, lawyers, physicians, clergymen and little country
gentlemen, who, having £2,000 or £3,000 at their command
to provide for younger sons, send them into the Army'.(92)
The majority of officers and aspirants however willingly
accepted these barriers on recruiting. Their constant
fear was that the recruitment of officers from a grade of
society to whom entry had been hitherto denied would mean
that 'men of a social rank much lower in the scale will be
able to present themselves as candidates for commissions. . . . The presence of these young men in the army
as commissioned officers may benefit the army, but the
possibility must be faced that it may not.'(93)   They
justified this attitude on the grounds that the lower
social status of such officers would unfavourably affect
their relationship with the rank and file. It was a
belief which persisted for a considerable length of time,
for it was a legacy of the neo-feudal concept that a soldier would more readily 'follow the gentleman', than an
officer whose social status was less certain and less well
assured.

But the erection of these barriers against 'open' recruiting into the Victorian military establishment must be
considered against the practice in other occupational
groups at this time. There is a tendency amongst critics
of the army to assume that the position in the military
was unique and that the introduction of open competition
had enabled men of ability to follow their chosen profession in any field other than the army. In practice, the
military establishment was not alone in setting up barriers against wide-scale recruitment, for in many nineteenth-century professions, particularly those which were
well-established, entry was expensive. In 1854, W.S.

Cookson in his evidence to the Inns of Court Commission
thought that £1,000 was the minimum outlay required for
entry into the junior branch of the legal profession.
Indeed, he thought that the total could be a great deal
higher when the cost of a premium at three hundred guineas
and the heavy stamp duty payable on taking up articles and
upon admission to the rolls was included. Nor was the
legal profession alone in imposing high charges designed
to 'increase the respectability of the Profession' by
keeping out the majority of aspirants. For architects,
premiums ranged from £100 to £500, so that the cost of
four years' articles could amount to £1,000, a figure
which was similar for the training of a civil engineer.
Professional education was not cheap, and the military
was very much in line with other occupational groups. Not
many families therefore could afford the cost of professional training, and an effect of this was that with some
exceptions, the majority of professionals, in the army and
in civil life, had to be drawn from a very small section
of society.
     In limiting therefore the area from which recruits to
the officer corps could be chosen the military, in this as
in so many other ways, was the mirror of the parent
society. The abolition of the purchase system had little
immediate effect on the overall composition of the Victorian military establishment, and the élite tried to
ensure a continuing control over recruitment, irrespective
of a political decision. There were indeed changes in the
background of recruits to the officer corps. After 1871
a decrease in the number of officers drawn from the families of the great landowners was balanced by an increase
in the number of entrants who came from the lesser gentry
and from the professional classes. But these changes did
not assume the proportions envisaged by the supporters of
open competition, nor did they weaken the preservation
and maintenance of a common military ethos, an ethos
firmly founded on the presumption that an officer was
without question also a gentleman. The military continued
to be an 'aristocratic' force, in which the presence of
dunderheads whose only claim to a commission was based on
their wealth and family background, was balanced by the
presence of men of ability such as Wolseley and his
colleagues in the Wolseley Ring. It is therefore impossible to evaluate the status of the Victorian military
establishment as a professional army solely in terms of
the expected changes which the critics of the purchase
system thought would follow on from its abolition.
Indeed after 1870, when entry was, in theory, more open,
there was evidence that changes in the pattern of

recruitment had retarded the development of professionalism, and to many critics, the Victorian military establishment remained, to its end, a non-professional army.(94)
Perhaps, after 1870, men of lesser means could obtain more
rapid promotion through intellectual ability, yet in 1904,
the Esher Committee found it almost impossible to find competent officers to fill senior appointments. 'We have
evolved a most logical and lucid system for administering
the Army,' commented Admiral Fisher, 'but Arnold-Foster
will tell you we are actually at our wits' end to find in
the whole of the British Army, suitable officers to fill
not only the few administrative posts in the new scheme
but for the Military Commands.'(95)

There was indeed a considerable amount of other superficial evidence to suggest that the army after 1870 was
far from being a professional army in the modern sense.
There appeared, for example, to be a continuing preoccupation with the niceties of Victorian social life
when, to enjoy this life, officers expected to get long
periods of leave which coincided with the demands of the
London season. Additionally, the officer corps in many
ways reflected in its habits and customs the norms and
mores of the society from which the majority of its members were drawn, while the wish of the remainder to be
imitative ensured that deviance from the expected pattern
of behaviour was exceptional.

In many ways, it was the image which the home-based
army projected in these closing years of the nineteenth
century that encouraged vehement criticism of the military.
The social structure of the officer corps differed to a
very limited extent from that of the early Victorian military. Within the closed world of the officers' mess a way
of life was maintained which reflected very closely the
leisured life of the landed interest. Horses and hounds,
fine crystal and expensive silver were symbols within the
regiment of social habits and attitudes which had been
unaffected by the reforms of the nineteenth century. The
customs and ritual of military life were to many external
observers archaic and outmoded, particularly when they
were seen against the background of the changes taking
place in society as a whole. It was the contrast between
the world of Samuel Smiles (1824-1904), retired secretary
of the South-Eastern Railway, with its attitudes reflected
in the titles of his books, 'Self-Help', 'Character',
'Duty', 'Thrift', 'Life and Labour', and the apparent in-
dolence of the military establishment. What other interpretation of military life could be placed on the expectation of officers that they would be allowed two days
each week for hunting?(96) How could an articulate middle

class, excluded from the officer corps by its expensive
standard of living and inadequate rate of reward, accept
that the military establishment was, after the abolition
of purchase, no different from any other institution in
society?

Equally the overseas army appeared to have altered very
little during the course of the century. The army which
invaded Afghanistan in 1839 had 40,000 camp followers. One
brigadier alone had some sixty camels to carry his personal effects.(97) But what difference was there in the
Sudan campaign of 1896? Steevens shows how the officers
attached to the Egyptian Army had overcome the hostility
of the desert:(98)

> I sat down at table and ate of soup and fish, of ragout
> and fresh mutton and game, and was invited to drink
> hock, claret, champagne, whisky, gin, lime-juice,
> ginger beer, Rosbach and cognac, or any combination or
> permutation of the same.

Outside the six-foot-high walls of Fort Atbara, only a
forest of stumps showed where the field of fire had been
cleared for over a mile in every direction. Inside, the
life of the regimental mess mirrored that of an English
country house even though the Fort was 1,200 miles from
the source of its supplies. In many permanent overseas
garrisons the standard of living was even more lavish than
at home. This was particularly noticeable in India where
officers, in common with their counterparts in the Indian
Civil Service, lived a deliberately ostentatious life.
At Lucknow in the 1890s, Smith-Dorrien and a fellow officer maintained between them some thirty horses and ponies
for racing and polo.(99) When Raja Pertals Regiment left
Skardu in the Kashmir to move south to Srinagar, 926
coolies formed the baggage train, thirty of them carrying
the baggage of the officer who had commanded them in their
campaign as part of the 1891 Hunza-Nagar Field Force.(100)
Consistently, so it appeared to the critics of the Victorian military establishment who were prepared to overlook the hardships of the North West Frontier, officers in
India lived a sybaritic life. Large private stables, long
periods of leave in Simla, the hill-station, expensive
messes and big-game shooting were considered to be characteristics which indicated how little the customs and
habits of the army had been affected by the 1870 reforms.

But any evaluation of military professionalism must
look beyond these characteristics, and consider more
critically the extent to which the Victorian army matched
up with the accepted criteria of professionalism. In this
context, the effect of the abolition of purchase should
not be minimized, but it must be evaluated together with

other factors. In retrospect, the arguments advanced by
reformers, such as Trevelyan, Higgins and De Fonblanque,
were at the same time both logical and irrational. They
expected that changes in the pattern of military recruitment would ipso facto alter attitudes towards the concept
of career and encourage the growth of professionalism.
Seen against the background of contemporary developments
in other occupational groups, this was a logical expectation. But it was an irrational aspiration in a situation
in which officers were 'content to be gentlemen', and to
leave the technical questions to 'those who were not
gentlemen'.(101) The question which arises is why the
officer corps in a period which revolutionized the nature
of warfare, armies and technology, continued to adopt a
dilettante attitude towards their career. Why did officers oppose all attempts to ensure the enforced professionalization of the military establishment? Attempts to
find a solution to this question are far from novel. It
was a question which equally intrigued Victorian critics
of the military system. To some, like Campbell-Bannerman,
this opposition could be attributed to the effects of
traditionalism. 'We are tied . . .', he told members of
the House of Commons, 'by traditions and prejudices and
habits which it is hopeless to overcome, and to ignore
which would be fatal to success. We have to humour the
feelings of those classes from which the Army is supplied.'(102) Yet these were the classes which supplied
the majority of members in the House of Commons, in the
professions and in the administration of the country. We
are still left with the question why it was that their
colleagues in the military retained attitudes and habits
which were outmoded even in Victorian society. To understand this, and to seek a possible solution, it is necessary to look more closely at the characteristics of Victorian military professionalism, and an explanation cannot
be attributed solely to the effects of the pattern of
recruitment.

4

◇◇

Professional Education

◇◇

Discussions of the characteristics of professionalism have been a recurrent theme in twentieth-century sociological writing from Flexner's statement in 1915 to the publication of Goode's work in 1969.(1) Most of the reached conclusions are, however, initially derived from an analysis of occupations whose professionalism developed primarily in the nineteenth century.(2) Clearly there are exceptions to this generalization in the case of occupations such as the Church and the Law which originated at a much earlier period, but even here, important professional characteristics were formally established during the Victorian period. Thus many of the universal characteristics which are used to describe professionalism today reflect the manner in which occupational groups had reached a particular stage of development by 1900. Of these, one characteristic in particular was indicative of the extent of professionalism in the Victorian army.

For the military, a core trait of professionalism was the educational and socialization process, since this not only directly produced the degree of expertise which was expected from group members, but it also affected elements of ideology, attitude and behaviour. Yet despite its importance, a preliminary judgment which can be made is that the standard of professional education in the Victorian military establishment was appallingly low. There are a large number of reasons for this, but an initial problem was that the army continually emphasized the importance of character as a criterion of recruitment, to the exclusion of intellect. The latter was considered to be of little immediate practical use, whereas the former could be seen to be the peculiar property of a traditional British élite, since it was a trait which reflected the finest virtues of the nation and the very foundations of the ethos of the state. Additionally, the emphasis which

the military establishment placed on 'character' was
derived from the nature of contemporary organizational
tasks. Garrison duties in colonial territories where the
routine of 'policing' demanded a knowledge of practical
subjects associated with quasi-permanent overseas postings
and the development of a pragmatic approach towards recurrent problems, reduced the demand for high intellectual
ability. It was moreover not only the military which came
to this conclusion. Sir William Denison, the Governor of
Madras, concluded in 1864 that what was needed in the
Indian Civil Service was not 'men of any extravagant amount
of knowledge. We would dispense with many of the "ologies",
if we could get with the ordinary education of a gentleman,
the habits and principles of a gentleman'.(3) These habits
and principles were in short, 'Their capacity to govern
others and control themselves, their aptitude for combining freedom with order, their love of healthy sports
and exercise.'(4) Essentially, these were qualities of
character which were not only needed in India where 'high
tone' was more important than efficiency,(5) but which were
equally important wherever the British were the rulers,
whether military or civilian. To be, for example, 'un-
afraid like the English' was looked upon by Steevens in
his jingoistic account of Kitchener's capture of Khartoum
in 1898 as one of the most important elements of the
English character. This lack of fear could be admired by
those whom England ruled:(6)

> And is it not good, ladies and gentlemen, as you walk
> in Piccadilly or the Mile End Road, that every one of
> these niggers honestly believes that to be English and
> to know fear are two things never heard of together.
> Utterly fearless themselves, savages brought up to
> think death in battle the natural lot of man, far pre-
> ferable to defeat or disgrace, they have lived with
> English officers and English sergeants through years of
> war and pestilence, and never seen any sign that these
> are not as contemptuous of death as themselves. They
> have seen many Englishmen die; they have never seen an
> Englishman show fear.

Equally, 'character' was an ideal attribute of the military establishment in England. Since one of the primary
tasks of the military was to preserve internal law and
order, this attribute emphasized the relationship between
the ruling class and the military, while it also confirmed
the claim of the latter that it was the 'sole and exclusive élite agency in society'.(7) In contrast, the criterion of 'intellect' was assumed within the army to be a concomitant of social and physical incompetence. It was
frequently equated with undesirable attributes which

implied that the intellectual was the boor and the prig
who knew books better than men, and who preferred work at
the desk to delights in the saddle. Translated into
practice, this attitude meant that no 'decent' regiment
during this period wanted to be burdened with men of an
inferior social standing whose only claim of entry into
the exclusive circle of the officers' mess, was that they
had acquired a sort of special knowledge for the fulfill-
ment of specific tasks. The military as a traditional
bureaucracy found it difficult to accommodate the conflict which the recruitment of such specialists engendered.
Whereas other occupational groups were able to permit a
great deal of internal strain and conflict, either potential or actual, the army, with its insistence on the need
for social homogeneity, was slow to accept that strain was
a natural and expected characteristic of large-scale
organizations. Instead it preferred to recruit officers
who would 'fit' in with existing group members, and whose
'character' traits could be expected to indicate a willingness to accept an established behavioural pattern, rather
than to gamble on the possibility that the intellectual
would question or reject expected norms and values.
   At the same time, there was little incentive in cavalry
and line regiments, where the level of technological development was low, for an individual who possessed intellectual ability to join the army. There were exceptions
to this generalization. Charles Kincaid-Lennox (1827-
1912), brother of Lord Bateman, was a Fellow of All Souls
who was also a Captain in the Life Guards, but the pattern
of purchased promotion gave officers such as these no
advantage over other less able officers. Even after the
abolition of purchase in 1871, there was little encouragement given to the individual officer to study for advancement or to develop his abilities, for the operation of the
seniority rule meant that men of intellect often took as
long a time as any other officer to reach field or general
rank. Moreover, the believed need to ensure the homogeneous nature of the establishment operated to discourage
the intellectual applicant to the officer corps. He was
always regarded as the oddity and the potential deviant.
The claimed merits of traits of character were thus preferred by the establishment as valid criteria of individual
acceptability, in the belief that these were a more accurate indicator of the potential loyalty of group members,
than any traits of intellect.
   In this situation, it was not surprising that there was
a positive military reaction towards the question of
theoretical knowledge. Although a body of systematic
theory serves as the basis of subsequent decision making,

and although the skills which characterize the professionalism of the occupation flow from this underlying
body of theory,(8) there was very little demand from the
Victorian army for the promulgation of such theory. The
association between the aims of the military education and
military goals was interpreted in such a way that 'theory'
was restricted in its meaning to refer primarily to tech-
niques of drill. Alternatively, it was equated with the
historical appreciation of military campaigns. Few officers cared to look beyond this narrow interpretation.
Consequently, the study of pure theory was subordinated
in importance as part of a training programme to the performance of drills, rituals and ceremonial activities
which supported the development of the trait of character.
All of these, moreover, were activities which enabled an
officer to identify himself with the corporate organization and to receive recognition as a member of the homogeneous group.

Fortunately for the professional development of the
Victorian army a small number of officers were prepared to
adopt a more liberal interpretation of 'theory'. Of
these, one of the most influential in the early part of
this period was Sir John Fox Burgoyne (1782-1871). After
Burgoyne, the eldest of four illegitimate children of
Lieutenant-General Rt Hon. John Burgoyne and Susan
Caulfield, the singer, went through Eton and Woolwich, he
began a brilliant military and civil career as an Engineer
officer. During his lifetime, he wrote copiously on
military theory, but his impact in this area was first
significant when he joined with Captain Charles Pasley to
form a group aimed at encouraging studies in 'the theoretical, practical and local in our complicated profession'.

Pasley (1780-1861) was another officer who enjoyed an
outstanding career in the public service. Born at Eskdale-
muir and educated at Selkirk where he was acknowledged to
be a brilliant scholar, he entered the RMA Woolwich in
1796. After service in Malta, Copenhagen, Spain and
Walcheren, he became in 1812 Director of the Royal Engineer
Establishment at Chatham, a post he held until his promotion to the rank of major-general in 1841. Subsequently,
he was Inspector-General of Railways before he retired
from public life. An impartial observer, Pasley was frequently called upon to give evidence to various Government
Railway Commissions, as in 1845 when he was the last wit-
ness called before the Gauge Commission which decided
between the rival claims of the broad and narrow gauge
English railway companies. Additionally, his specialist
advice was often sought by railway contractors who were
faced with a particularly difficult engineering problem.

When Cubitt, for example, was constructing the South
Eastern Railway between Folkestone and Dover, his plan to
demolish completely Round Down Cliffe, 375 feet above sea
level, was discussed by him with Pasley. On the latter's
advice, the actual demolition was planned by Lieutenant
Hutchinson, RE, one of the few men in Britain at this time
who was experienced in the simultaneous firing of large
multiple charges. Elected a Fellow of the Royal Society
in 1816, Pasley published a number of works, some of which
dealt with various aspects of military theory — 'Essays
on the Military Policy and Institutions of the British
Empire' (1810), a 'Manual of Military Instruction' (1814)
and 'Rules for Conducting the Operation of a Siege' (1829).
Additionally, Pasley's inventive genius led him to propose a reformed and uniform system of Imperial weights and
measures (1834), to point out the advantages to the construction industry of using artificial cement (1836) and
to conduct a series of experiments in the removal of
under-water obstacles (1836). In 1810, though, at the
start of their careers, Burgoyne and Pasley were more
concerned with their Society for Producing Useful Military
Information. This comprised six charter members, all of
whom were Engineer officers then engaged in operations in
the Peninsula.(9) While the society did not survive the
end of the Napoleonic Wars, Burgoyne in particular continued to show a degree of interest, unusual in an officer
of this period, in the study of military theory. As a
theorist he was considerably influenced by the Swiss
writer on military affairs, Baron Jomini, whose classic
work, 'Traité des grandes operations militaires' was first
published in 1811. Burgoyne prepared a detailed appreciation of this work, but his comments never appeared in
print, and it was left to another army officer to bring
this outstanding Continental work before a British military
audience. In 1821, a review of Jomini's book was the first
publication of Major-General Sir William Napier,(10)
(1795-1860). An artilleryman who had also served with
Moore in the Peninsula, Napier went onto half pay as a
major in 1819 to devote his life to the arts. Not only
was he elected as an honorary member of the Royal Academy
for his skill as a sculptor, but his monumental study,
'History of War in the Peninsula', which was a defence of
Sir John Moore, was published by John Murray in London in
1828.
     The numerous writings of Napier and Burgoyne on a wide
range of theoretical and practical military subjects were,
however, rivalled by those of Major-General John Mitchell
(1785-1859), a Scotsman who was the son of John Mitchell
of the Diplomatic Corps. Educated at Ritter Academy,

Luneburgh and Mr Nicholson's mathematical school in
London, Mitchell was renowned for his linguistic ability,
a gift which brought him an appointment on Wellington's
Staff in Paris after the end of the Napoleonic Wars. A
major in 1821 after eighteen years' service, Mitchell went
onto half pay in 1826 and devoted himself to writing. A
contributor to many contemporary journals under a variety
of names, Mitchell was best known for his 'Life of
Wallenstein' (1837), his 'Thoughts on Tactics', and for
his contributions to the 'United Services Journal' from
1841 to 1855. Indeed of the four theorists whose writings
spanned the years from 1821 to 1870, it was Mitchell who
was most aware of the association between military professionalism and the study of military theory. Napier
was primarily a strategist and historian who developed a
thesis of the relationship between historical events and
the accepted principles of war. Like Pasley, his attitude
towards many contemporary army problems were those of the
traditionalist, and his views on military affairs reflected an interest in preserving much of the existing system.
Burgoyne, as a military engineer, was increasingly concerned with the impact of a developing technology upon
traditional methods of warfare, but his innate conservatism
was revealed in his attitude towards such subjects as educational tests for officers, his rejection of promotion by
merit and his dislike of a separate Staff corps.(11)

In contrast with these officers, all of whom developed
theories within the constraints of the existing system,
Mitchell considered that there was a positive relationship
between the professionalization of the military and the
extent to which military theory was studied. To him, the
low degree of military professionalism at this time was
both a cause and effect of the limited appreciation
possessed by officers of the importance of theory:(12)

Officers enter the army at an age when they are more
likely to take up existing opinions, than to form
opinions of their own. They grow up carrying into
effect orders and regulations founded on those received opinions; they become, in some measure, identified with existing views till in the course of years,
the ideas thus gradually imbibed get too firmly rooted
to be either shaken or eradicated by the force of argument or reflection. No sooner is an officer looked
upon as a theorist and innovator than he is set down
as an unhappy person.

In this summary of the treatment which was meted out to
the non-conformists, Mitchell thus recognized the defects
of a system in which there were, as one critic said,
'Superiors always more ready to check, than to forward,

the advancement of inferior officers possessing higher
attainments.'(13)

   With the exception of the writings of these theorists,
however, little else was published during the early years
of the Victorian army which contributed to the development
of an underlying body of theory. This dearth of theory
was also noticeable in the USA. In a catalogue of 1856,
the majority of published works covered the three areas of
military law and administration, military history and corps
drill. Of the three books on the 'Art of War' one was an
1854 translation of Jomini by Major O.F. Winship, Assistant
Adjutant-General, United States Army and Lieutenant E.E.
McLean, 1st Infantry, United States Army. The second,
'The Art of War', was an 1815 reprint edited by Robert
Fulton. The third, 'Elements of Military Art and Science'
by H. Wager Halleck, United States Army, was published in
1846. The general attitude adopted by most serving officers was consequently very clear:(14)

   Of formal books on tactics, whether by such are meant
   didactic essays or fanciful theories, our service has
   fortunately produced few . . . we believe that more
   sound knowledge of the principles of our profession is
   to be gained by the attentive perusal of a few pages
   of works such as those of Jones or Pasley, Napier or
   Douglas, than from the study of whole volumes of professed tactical essays . . . the whole subject of the
   'Grande Tactique' will never be taught or learnt from
   mere books.

   Yet in accepting that during this period there was an
almost complete absence of a systematic body of military
theory in Britain, it is important to relate this deficiency to its historical context. The acquisition by
potential professionals of this theoretical base depended
in any occupation not only on the output of recognized
theorists but also on the encouragement given them by
group members. In this respect the typical attitude of
the military officer towards the need for courses of professional education which contained a high theoretical
content, mirrored the attitude of his contemporaries in a
wide number of occupations. In both cases, military and
civilian, attitudes of individuals were primarily conditioned by their previous educational experience.
Opposition to professional studies, therefore, reflected
a latent or manifest appreciation of the alleged differences between the aims of professional education and
the purpose of higher or liberal education. This difference was consistently brought out by such eminent
Victorian educationalists as Arnold, Butler and Thring,
who defined education as training for leadership, not

tuition for scholarship, an attitude which led them to
separate preparation for the universities from vocational
training for the middle class.(15) As Thring commented
'if professional education is the object, then by all
means establish schools to give it. But they cannot be
great schools that train for the Universities, for they
go on the principle that education is the object.'(16)

The established education system was thus designed to
promote instruction in a narrowly restricted range of
subjects. In law, the 701 endowed grammar schools which
were in existence were obliged to teach the classical
languages — Greek and Latin — and little else. Indeed
in 1805, and again in 1826, Eldon, the Lord Chancellor,
had ruled that it was 'ultra vires' for a grammar school
to use its endowment to further any other educational
aim.(17) As a result, it was initially left to a number
of poorly endowed grammar schools, and a number of private
schools which ranged in their efficiency from 'Dame
Schools' to 'Private Academies' to teach subjects other
than the 'business of the classics', and to meet the needs
of the middle class with regard to professional training
based on scientific instruction. Unfettered by ancient
statutes or traditional methods of instruction, these
private institutions, many of which arose from middle-
class opposition to the network of endowed schools, taught
an extended curriculum which included reading, writing,
arithmetic, history, English grammar, geography, the rudiments of physical science and, in a few cases, land-
surveying or book-keeping.(18) In a small number of these
private academies, some 200 of which were in existence, a
positive attempt was made to introduce the teaching of
subjects which were believed to be useful for aspiring
members of the officer corps.(19) These schools were,
however, few in number with doubtful standards of attainment, and their status as educational institutions was
far below that of the traditional classical school.

By the time of Queen Victoria's accession to the throne,
however, significant changes were taking place in the
sphere of education. A fourth type of institution, the
proprietary school, had been added to the categories of
schools listed above, while a number of reforms had been
introduced at some of the endowed foundations. These
changes were not autonomous. They were the result of
middle class pressure to have access to an educational
system compatible with their aspirations.(20) The
struggle, as has been pointed out elsewhere, had a two-
fold aspect:(21)

It involved, first, a sharp ideological attack against
existing institutions, and, with this, the formulation

of a positive alternative educational policy both as regards form and content; second, the establishment, by the middle class, of its own institutions under its own control, serving its own purposes, and untrammelled by ancient statutes and clerical control.

The proprietary schools thus marked a transition from the mere seeking out of schools which promised a relevant education, to the foundation and control over schools of the type desired. This development reflected not only a demand for systematic education, but also for more stable educational forms. At first the proprietary schools tended to be based in larger towns and cities, but with the growth of rail transport and the quasi-respectability afforded boarding schools by Arnold's reforms at Rugby, proprietary boarding schools located in the country became an increasingly attractive proposition. In response to specific middle-class needs, they began to multiply during the 1840s and 1850s. Marlborough College (1843) was representative of this development, for it was founded for the sons of the poorer professional classes and clergy. Rossall (1844), similarly, was for the sons of the clergy and laymen of modest means. The Woodard Foundation Schools, though an endowed foundation, were also a part of this movement, since they were designed deliberately to meet the requirements of all sections of the middle classes.

The majority of potential officers, however, attended the traditional endowed schools, some of which were already receiving recognition as 'public' schools, one of the major sources of officer recruitment in the later Victorian period. In these boarding schools, they joined with the majority of the British élite in receiving a common classical education which posited a connection between the defence of classicism and social leadership. This had been brought out in 1820:(22)

> The rich and those, who, if not rich, yet enjoy a competency have parts to act in society, which the poor are neither required nor able to perform. To qualify them for these parts, the rich usually have the means of a more comprehensive education; that sort of education, which is termed liberal, and which is accorded at the grammar school.

Few attended the non-grammar schools, described by Knox as 'schools for the shop, the warehouse, the counting house and the manufactory'. Indeed, from a sample of 3,000 names drawn from the 'Dictionary of National Biography', Hans calculates that nine-tenths of the élite in Britain were educated at the classical schools.(23) One result of this was that the early education of most

officers differed completely from that of entrants to
other professions. The latter argued that concentration
on the classical disciplines had become unsuited to the
perceived needs of the middle classes, the majority of
whom would be destined for the world of business, commerce
or the newer professions, rather than the universities.
Their demand was for a more Utilitarian form of education,
of the kind suggested by Jeremy Bentham, in which the provision of tuition and subject-matter was in accordance
with vocational requirements.

Officers, in contrast, continued to enjoy a narrowly-
based education. Even those who went on to university
before joining the army could not escape from public insistence on the importance to the individual and to society
of a good classical education. Samuel Butler, for example,
bitterly criticized Cambridge University for spending too
much time on the teaching of mathematics rather than on
the classic tripos. While neither Cardinal Newman nor
Pusey can be considered fully representative of Oxford
thought in these early Victorian years, Newman was confirming a generally accepted view when he argued that,
'The problem and special worth of our university is not
how to advance science, not how to make discoveries . . .
but to form minds, religiously, morally, intellectually. . .
It would be a perversion of our institutions to turn the
university into a forcing house of intellect.'(24) The
whole purpose of a higher liberal education was thus considered to be the advancement of those branches of knowledge to which it was desirable an English gentleman
should pay some attention.(25) It was firmly believed
that it was not the purpose of the public school or university to provide a vocationally-oriented education, a
sentiment which Thomas Arnold expressed very clearly in
1836. 'Surely the one thing needed for a Christian and
an Englishman to study is a Christian and moral and
political philosophy.'(26) For those students who did
contemplate a professional career, the strength of their
previous school experience was that it gave them the
classical education which, it was presumed, every other
entrant would have enjoyed. If there was a need to acquire
technical knowledge, then it was accepted that this was
most readily acquired through some form of apprenticeship
or pupillage. The essential objectives of the public
school and the university were thus the transmission of a
body of central cultural values, to the total exclusion
of considerations about vocational or professional needs.

In its early years, the Victorian army therefore faced
a number of problems which had to be overcome if the education of officers were to be improved. In common with

other occupational groups, attempts to develop professional knowledge were hindered by the legacy of the classical tradition and the concomitant effects of a preference for character rather than intellect. The effect of this was succinctly summed up by a contemporary commentator:(27)

> Thus Arnold subordinates mere knowledge which is termed professional training, to the true formative education, which is termed liberal. The humanities, and particularly the study of the classics, are the most formative subjects, when taught in a Christian perspective. It may well be doubted whether we would ever have regarded any acquaintance with the material forces of nature as good substitutes for the intellectual culture derived from classical studies, or as equal to them in disciplinal value.

Yet in trying to escape from the constraints imposed by the pattern of an English gentleman's education, the army, in comparison with other occupational groups, did enjoy the advantage of possessing its own academic institutions. Of these, the oldest, the Royal Military Academy at Woolwich, had been specifically established to meet the demands for formal education born out of the technical skills required in the Scientific Corps of Artillery and Engineers. These demands were not peculiar to Britain. In France, Russia and Prussia, the first schools of officer training were those designed to meet the needs of artillerymen and engineers. In Britain, moreover, this need had been recognized as early as the mid-eighteenth century when by a Royal Warrant the Royal Academy, later the Royal Military Academy, was set up, since,(28)

> It would conduce to the good of our service if an Academy or school were instituted for instructing the raw and inexperienced people belonging to the Military Branch of this office in the several parts of mathematics necessary to qualify them for the service of the Artillery and the business of Engineers.

The emphasis placed initially on instruction in mathematics, normally studied only in the universities,(29) was gradually extended until Woolwich was one of the few institutions of higher education in the United Kingdom which offered a technologically-orientated course. During a period when the scientific revolution had scarcely influenced the English universities,(30) and when a contemporary commentator could allege that 'within the last fifteen years not a single discovery or invention of prominent interest has been made in our colleges',(31) the Academy seemed to be in a particularly favoured position. Herbert Spencer, writing in 1861, subsequently drew attention to this persistent failure of the English

educational system to meet the demands of everyday
life:(32)

> And here we see most distinctly the vice of our educa-
> tional system . . . it neglects the plant for the sake
> of the flower. In anxiety for elegance, it forgets
> substance, while it gives no knowledge conducive to
> self-preservation. While of knowledge that facilitates
> gaining a likelihood it gives but the rudiments, and
> leaves the greater part to be picked up anyhow in after
> life.

But in contrast with this general criticism of contemporary
civilian education, the Royal Military Academy, in theory,
appeared to ensure that potential officers in the Artillery
and Engineers were participants in an educational process
which was directly related to the development of a highly
professionalized occupational group. A planned programme
governed the entry training and certification of the student in the qualifications required on entry. Gentlemen
Cadets were expected on entry to possess a knowledge of
English, Mathematics, Latin, French, Geography, History
and Drawing, a range of subjects far wider than that associated with a classical education. In addition, since
there were four nominations for every three vacancies, a
preference could be shown for applicants who had a further
knowledge of German, Greek and Advanced Drawing.(33)   In
the training process, theoretical studies taught by
civilian professors of high academic standing were complemented by courses of practical application given by
military specialists. Passing-out examination seemed to
ensure that a reasonable standard of attainment was
achieved by those who graduated, and a failure rate of
a quarter during the years from 1825-49 suggested that
those who failed to reach this standard were rigorously
eliminated.(34)

Yet in practice the standard of professionalism among
members of this group was disappointingly low. There
were some notable exceptions, but in general terms the
Academy did not fully realize its potential even though a
number of graduates subsequently occupied distinguished
civil appointments. There were a number of reasons for
this. From the beginning, the age of entry into Woolwich
was set at too low a level. The original age limits on
entry of fourteen to sixteen had been raised by the revised
regulations of 1835 to fifteen to seventeen,(35) but it is
evident from contemporary military returns that a number
of the candidates were younger than this on entry. The
'Return of the Royal Military Academy dated April, 1856',
for example, shows that in 1821 the average age of the
five cadets admitted was fifteen years two months. This

decreased to 14.6 in 1831, rose to 15.8 for the cadets
admitted in 1841, but decreased again to 14.11 in 1851
and to fourteen years eight months for the ninety-six
cadets who were admitted in 1855. Much of the theoretical
studies, particularly during the first year of the course
prior to the probationary examination in Mathematics and
Fortifications, was therefore orientated towards meeting
the needs of these youngsters. The curriculum was accordingly designed to educate schoolboys, not to meet the
requirements of aspiring military officers. The effect of
this was especially noted by the civilian teaching staff
who, finding that they were being forced to devote their
time to instruction in basic principles, urged that, 'The
age of admission should be considerably raised. My experience in the Academy has left me under the painful
impression that the Cadets are just beginning to take an
interest in their studies and to comprehend their scope,
when they are about to quit them, too often never to be
resumed.'(36)
    Senior Ordnance Officers who had taken part in the edu-
cational training programme confirmed this point of view.
As Major-General W.D. Jones of the Royal Artillery commented, 'The time of the masters, for the first year, is
occupied in imparting the very rudiments and elementary
knowledge of the subjects they [the Cadets] ought to have
soundly known in joining, which leads to so many being
removed at the probationary examinations.'(37) The precise number of cadets who were removed from the Academy
at the Probationary examination stage, was, however, less
than Jones implies. Marks gained by cadets in the two
subjects of this examination were persistently low throughout this period and few candidates obtained marks
approaching the maximum number obtainable, but few cadets
were discharged. From 1842 to 1850 (inclusive) 50 cadets
out of the 627 original entrants were removed. In 1841,
however, the figures support Jones's contention, for 20
cadets from an entry of 88 were discharged. It is noticeable that in the years when the average age on entry rose
above sixteen, as in 1845 and 1846, the number who were
discharged decreased considerably, in this case to one and
none respectively. But the emphasis which was placed on
the importance of these Probationary examinations reflected
the initial defect in the Woolwich system. It was not
until 1857 that entry to the ordinary course at the Academy was by competitive examination, and for as long as a
system of nomination persisted, it was perhaps inevitable
that boys who entered Woolwich at the age of fourteen
entered a militarized 'public school' rather than a
military 'college'.

The actual curriculum of the Academy in these early years of the Victorian military establishment was, in theory, designed to further a study of those subjects which were related to the tasks carried out by officers of the Scientific Corps. But it is questionable how far the studied subjects contributed to the development of professionalism. The emphasis placed on the study of mathematics, for example, produced a paradoxical situation. On the one hand, there was a need for the study of this subject to a greater depth than that attained on the training course if any reasonable standard of proficiency were to be reached. Conversely, the existing emphasis precluded the study of subjects such as 'regimental duties' which were more directly related to the subsequent tasks carried out by the specialist officer. Here, there was a conflict between the needs of theory and the needs of practice, and, in trying to effect a compromise be- tween conflicting educational demands, the Academy failed to develop either to the full. The ensuing unsatisfactory situation was then exacerbated by inadequate feedback from theoretical to practical classes, so that academic know- ledge gained in the former was not reinforced in the much shorter time devoted to the study of the practical appli- cation of acquired knowledge.

The methods of instruction, in addition, did little to motivate individual interest in academic study. The sub- jects studied were, to a very considerable extent, 'new' subjects or were exclusive to the army, so that, in the absence of a developed body of theory, there was a marked shortage of suitable text books. Cadets made voluminous notes from the lessons given them and carefully copied out manuscripts and drawings, but since the majority of civilian staff were continuously employed at the Academy for upwards of thirty years, lectures became increasingly out-dated as a form of 'rote learning', and it is evident that much of the transmitted information failed to take into account the technological developments which were taking place in Victorian England.

While a complex system of final examinations had been evolved, their significance as indicators of a reached level of professionalism, was in fact reduced by their use as a bureaucratic selection process. These passing-out examinations had been regularly held from 1764 to 1793, and although they were dropped during the Napoleonic Wars to meet the immediate needs of an expanding army, they were reintroduced in 1811. After this date, they were continually used to determine the branch of the Scientific Corps into which the officer was allocated on graduation. Successful candidates were allowed to choose between the

Artillery and the Engineers according to their ranked position in the final examination, until all the vacancies in
a particular corps were filled. Thereafter, the remainder
of the graduating class were automatically allocated to
unfilled vacancies. In practice, the smaller number of
vacancies in the Engineers was taken up by cadets at the
head of the list of merit. Other less successful cadets,
irrespective of their ability or their suitability for the
appointment, were drafted into the Artillery. The 'Return
of the Royal Military Academy dated 26th April 1856', cited
as Appendix F of the 'Report on Training Officers for the
Scientific Corps' (1856) shows that in the thirty-six final
examinations held between June 1841 and December 1855, 525
cadets were commissioned in the Artillery and 205 in the
Engineers. On thirty-one occasions, the cadet at the head
of the list of merit chose to enter the Royal Engineers.
In twenty-three examinations, his example was followed by
the top third of the list of merit. Possible success in
the operational role was, therefore, related neither to
preference nor to achievement, and compulsory allocation
to a corps not of his choice, did little to encourage the
officer in developing further the theoretical knowledge
gained at the Academy.
   In addition, the form of the final examination was
defective, because its importance as an educational test
was consistently reduced by the need to ensure a constant
flow into the corps of new entrants. The shortage of
Scientific Corps officers was often most marked. Indeed,
it was the inability of the Academy to meet the demand for
officers during the Crimean War which led to the introduction of the revised entrance requirements. Equally, the
shortage of officers affected rates of graduation and the
effect of this was noted by a number of contemporary
critics. Colonel Sandham, the Director of the Royal
Engineers Establishment at Chatham in 1857, complained
that 'The junior officers . . . are always at a loss to
apply their knowledge; the accuracy of their knowledge,
therefore, might be questioned. They are generally deficient in drawing: they should have more practice and
instruction of every description of drawing.' Professor
J.J. Sylvester was less specific in his criticism and
contented himself with saying that 'the attainment of the
cadets on leaving the Academy . . . are insufficient,
meagre and unsatisfactory'. A similar comment was passed
by Lieutenant-Colonel M'Kerlie, who, as a member of the
Irish Board of Works, received many newly-graduated officers as additions to his establishment: 'As regards the
great number, though the knowledge may be considerable in
extent it is nevertheless but superficial, and as a matter
of course is soon forgotten.'(38)

In part, these deficiencies might have been overcome
by prolonging the training process to ensure a more comprehensive coverage of requisite subjects. Alternatively,
or in addition, a programme of post-graduate courses could
have ensured that officers received an opportunity to participate in a programme of further education and training.
As it was, the length and scope of professional training
was restricted to the period of initial instruction at the
Academy, so that few officers, even if they wished to
continue with their studies, were given the chance to
keep abreast of technological developments. The problems
to which this could give rise were dealt with at some
length in the 1838 Commission of Enquiry into Military
Promotion. In a series of answers to the Commissioners'
Questions, Major-General Frederick Thackeray, the senior
representative of the Royal Engineers in Ireland, commented upon the attitude of officers under his command.
Ireland was a particular case, in the sense that Engineer
officers were employed mainly in support of the civil
administration, so that their duties were wider in scope
than was normal, but Thackeray's comments echoed the
feelings of many other members of the military establishment:

Question 962
'Do you mean to say that the education at the
Academy at Woolwich is not at this moment as good an
education as that in any other military academy that
you are acquainted with?'
Reply
'Not knowing precisely what the course of instruction at Woolwich now is, I can hardly answer that
question. . . . The officers under my command in
Ireland were almost entirely employed very much in the
details of barrack life. We had a professional library
under the roof of my own office and I found that I had
no time to read any professional books, and I asked
the gentleman who was the librarian whether the officers took out books, and he said that a few took out
books, but not professional books.'
Question 964
'Then do you consider that an officer of engineers,
having quitted Chatham and having since been employed
for some years in the charge of barracks, is not competent to go into the field and fulfil the duties that
may be expected of him?'
Reply
'He is not so competent, as he would have been had
he been studying his profession.'

Question 965
    'How would you employ him in the study of his pro-
fession?'
Reply
    'I would give him leisure to read military works of
fortifications and on strategy.'
Question 966
    'Do you mean to say that an officer had no leisure
or no opportunity of making himself acquainted with his
profession after he has quitted the Academy and entered
upon his duties as an engineer officer?'
Reply
    'Those in Ireland had very little.'
Question 969
    'Looking at it from a public point of view, do you
conceive that the public would allow officers of
engineers, in time of peace, on the plea of pursuing
their studies . . . to remain apparently idle for the
sake of obtaining theoretical information?'
Reply
    'Certainly not.'
The extent to which programmes of further education and
re-training should be provided for officers of the Artillery and engineers was also debated by the 1856 Commission
on Training Officers for the Scientific Corps. The question which was posed was quite specific:(39)
    'What instruction or assistance in their professional
    studies is at present supplied to Officers in the
    Artillery or Engineers, after joining their respective
    Corps?'
This was submitted to sixteen Artillery officers, representing officers of all ranks from General to Lieutenant,
and to fourteen Engineer officers of comparable seniority.
Of the Artillery officers who gave evidence to the Commission, all were agreed that apart from the six months'
course given to direct entrants from civilian life who had
been commissioned to meet the needs of the Crimean War,
there were no facilities for providing organized instruction or assistance for officers. Some, like Major
Charteris, RA, considered that this was quite satisfactory:
    'When officers have received their Commissions in the
    Corps or Artillery, their previous education at Wool-
    wich Academy and the examinations they have undergone,
    ought to ensure the possession of a considerable amount
    of theoretical and practical knowledge of their pro-
    fession.'
The majority, however, rejected this narrow interpretation
of the effects of the initial training programme, and
General Sir Hew Ross went so far as to propose a 'corres-

pondence course', to improve the standard of military
education:

The Royal Artillery Institution with the powers of
extension and correspondence through its Secretary,
and with a system of detached professional libraries,
which could be easily established at the out-stations,
would afford the means of carrying on professional
studies and improvement.

It was perhaps significant that the most damning
criticism of the existing system of British military
education in this Crimean period came from the senior
officer to give evidence before the commission. General
Sir Howard Douglas, who had made an extensive study of the
military education undertaken by young officers on the
Continent, was very searching in his comments on the
situation in the United Kingdom:(40)

The entire subject of Military Education should be
revised and made to accord with the present organization of the army. . . . Most young men, after joining
the Regiment or Corps, neglect to keep up with the
knowledge they have acquired at the Academy, and are
thus incapable of applying it to useful purposes. The
few officers of Artillery and Engineers who have distinguished themselves in the scientific world are
bright exceptions.

Although Douglas's criticism was directed specifically
at the Scientific Corps, his comments had a wider applicability, for while graduates of the Academy enjoyed a certain professional reputation, a low level of professionalism was even more evident in the remainder of the Victorian
army. In this much larger group criticism of professional
standards had a lengthy historical background. The Duke
of Wellington, for example, had complained during the
Napoleonic Wars of the low educational standard, want of
military efficiency and poor powers of leadership possessed
by his regimental officers. There were, however, fundamental differences in the respective relationship between
planned courses of training and the development of military
professionalism, since the needs of the two groups were
quite distinct. For the Scientific Corps, the essential
problem was the difficulty of reconciling the objectives
of military education with the impact of the industrial
revolution. For the army, that is the cavalry, the Guards
and the line regiments, any attempted development of occupational professionalism generated a host of complex
problems, Initially, it was the difficulty of striking a
balance between the needs of the officer as a member of a
homogeneous social group, and his needs as a member of the
army in an era of change. Group requirements, derived

from the continuing existence of a neo-feudal attitude of
mind, were widely based. In Britain, as in Prussia, these
requirements were essentially normative in origin:(41)

> Education and technical knowledge are not the only
> things required for a competent officer; he needs
> also presence of mind, rapid judgement, punctuality,
> regular habits on duty and proper behaviour. These are
> the cardinal virtues that every officer must have.

In contrast, the needs of the army demanded a specialist
education and training which were related more positively
to the development of technical knowledge and expertise.
The regimental emphasis on 'character' was thus balanced
by the army stress on 'intellect' as contrasting criteria
of efficiency, but any programme designed to meet either
or both of these aims was further affected by the existence
within the military, as in other occupations, of a spirit
of Utilitarianism. While the number of officers who advocated this approach to military training was small, their
arguments reflected to a certain degree contemporary
civilian opinion. The educational philosophy which could
be associated with Utilitarianism was to some critics of
the army attractive, since it negated the connection between ascription and intelligence, and therefore between
birth and the right to higher education. This implied
that the traditional educational system which excluded much
of the middle class from secondary and higher education was
irrational and indefensible, so that the Utilitarian philosophy epitomized the ambitions of an emergent middle class.

In relation to military education, Utilitarianism suggested two avenues of development which attacked the existing established pattern. On the one hand, its rejection of
an ascriptive-based education concomitantly implied the rejection of any programme designed to meet the needs of a
homogeneous élitist group. On the other hand, this philosophy suggested that the content of instruction should be
defined in relation to military function and that all military education should be centered upon vocational training.
The problem which persisted was that the majority of officers preferred the amateur traditions of the public school
system, having no wish to see this replaced by a programme
of useful instruction allegedly based on the claim that
military education and training should be available to the
middle as well as to the upper class. Professionalism, because of its association with Utilitarianism seemed to many
of these objectors to be a middle class phenomenon, the
adoption of which threatened external recognition of the
army as an élitist institution. Thus in rejecting professionalism and in denigrating courses designed to promote
it, officers' attitudes were conditioned by their apprecia-

tion of 'tone' and 'status' and their dislike of the
Utilitarian wish to reduce social inequality. Even so,
for the supporters of this philosophy there appeared to
be no reason why this principle could not be adopted in
the military:(42)

It would be a simple process to correct the educational
process and make it perfect, providing the utilitarian
principle, which forms part of the national character
and governs most of our institutions, be fully adopted
in the military establishment.

As at Woolwich, army education and training were primarily associated with the existence of an educational
institution. The Royal Military College had been opened
in 1802 at Great Marlow, 'to provide an education for the
sons of officers who were intended to follow their fathers'
profession and to enter the army'. This early attempt to
create a self-perpetuating group, was, however, short-
lived, for the creation of a 'stratocracy' was obnoxious
to a civilian society which equated this with Cromwellian
militarism, and after 1803, the College admitted young
Gentlemen Cadets, aged between thirteen and fifteen, who
came from a more varied social background.(43) By the
beginning of the Victorian period, a planned programme
comparable with that of the Royal Military Academy at
Woolwich, governed the entry, training and graduation of
the students. An aspirant was nominated by the Commander-
in-Chief. The candidate sat a simple qualifying examination, the standard of which was lower than the comparable
examination sat by nominees for Woolwich. Once admitted,
the cadet undertook a course which emphasized the study of
mathematics, fortifications and military drawing before he
sat a final examination.

In these early years, little of this course did much to
develop military professionalism. As at Woolwich, far
too much time was spent on remedial education, supple-
menting the deficiencies of the general education which
cadets had received in their 'classical' schools. Concomitantly, the syllabus covered too wide a field. In
the final examination, for instance, Sandhurst cadets had
to satisfy the Inspector of Studies and the Board of Commissioners in the 'six steps essential to obtain a commission — mathematics, fortification, military surveying and
three optional subjects chosen from French, German, siege
operation, landscape drawing, military drawing, Latin,
general history and geography'.(44)

The list was impressive but as critics pointed out there
was no study of the more practical subjects which would
have been useful to a young officer, that is, subjects
such as military law, administration, logistics, transport,

communications and hygiene. It is also evident from the
example of foreign military academies that the theoretical
content of the Sandhurst course did not measure up to that
of courses provided abroad. Thus, while Sylvanus Thayer
at West Point evolved a curriculum designed to produce
officers who could participate fully in the decision-making
processes within a society,(45) the Sandhurst programme
tried to meet too many aims — general education, military
education, military training — and failed to satisfy any
single one.

These deficiencies in the curriculum of the Royal
Military College were, to a very large extent, the result
of the College only partially realizing the aims of its
founder. In formulating a planned course of military
education which would 'replace the power of nepotism which
for so long had stifled talent',(46) the originator of the
plan, Colonel Le Marchant (1766-1812), a renowned cavalryman and author, envisaged three distinct courses of study
in separate departments. The senior course was to give
training in staff duties and in the theory of war to selected regimental officers. A second was to cater for
aspiring ensigns who were to be instructed in military
subjects, horsemanship and sword-drill for six months before they joined their regiments. The third course was
designed to give a thorough grounding in 'scientific subjects' to young boys of thirteen to fifteen who intended
to seek commissions in the army. When the Royal Military
College was established, however, the junior department
was an amalgam of Le Marchant's second and third courses.
It strove to implement the aims of the second but with
cadets whose age range was more appropriate to the proposed third department. Because of the educational shortcomings of these young cadets, the College was forced to
provide instruction in subjects appropriate to the remedial
needs of schoolboys, as well as in those subjects which
were required by the aspiring officer. The result was to
create a dichotomy of purpose. In part, this was resolved
by over-emphasizing the importance of 'school' subjects,
to the extent that the Prince Consort could later complain,
'What is to be gained by making officers of the army . . .
abstract mathematicians instead of scientific soldiers. . .
it is a well ascertained fact . . . that mathematicians
from their peculiar bent of mind, do of all men show the
least judgement for the practical purposes of life.'(47)

This uncertainty of purpose was to bedevil Sandhurst
for many years until the post-Crimean reforms. Essentially,
the College had, in view of the youthfulness of its entrants, to serve as a public school. Courses were therefore designed for boys who had only partially completed

their education. Yet in recognizing this defect, any
attempt to modify the system was inevitably affected by
the knowledge that in reality only a small number of
officers did go through Sandhurst. This suggested that
little importance could be attached to a Sandhurst training as a qualification for recruitment and subsequent
promotion. As far as recruitment was concerned, the
position in the British Army was in accord with European
thought where general opinion, as a whole, was opposed to
the idea that all officers should pass through a Military
College prior to commissioning. Even so, the numbers of
newly-commissioned officers who had passed through Sandhurst was low by any standard. In the four years from
1834 to 1838, for example, the College provided less than
one-fifth of all new entrants into the army.(48) This
figure varied in subsequent years, but Sandhurst officers
were always a very small minority, and it was noticeable
when the army suddenly expanded in size, as in the Crimean
War and after, that an increasing number of officers were
appointed directly from civil life. For these officers,
no attempt was made to give them instruction in even the
basic military subjects, so that up to four-fifths of the
officer corps reported straight to their regiments on
commissioning, where they tried to acquire military expertise through some form of on-the-job training.

Some attempts were made to ensure that these non-
Sandhurst officers had attained a basic standard of education. In 1849, for example, the Duke of Wellington, partly
in response to public pressure, issued new regulations for
candidates who wished to obtain a commission.(49) Three
years previously, Lord Grey in a written memorandum on
military education had complained that there was no guarantee that newly-appointed officers were competent to dis-
charge their duties. He had therefore urged the Cabinet
to introduce an examination for first commissions and for
promotion from the rank of captain to major.(50) Now to
meet these complaints, Wellington promulgated instructions
which stated that no one could be commissioned 'unless he
could prove by examination to have good abilities and to
have received the education of a gentleman'. In addition,
no one could be promoted to lieutenant or captain, 'unless
he could satisfy a tribunal of his professional and general
acquirements and fitness'.(51)

These regulations only partially solved the problem.
The army, finding that the examinations proved too difficult for candidates, adjusted the qualifying standard. To
make them more suitable for candidates from public schools,
the balance between the subjects had to be altered to
reflect more closely the curriculum of these schools. Once

again the wish to ensure social homogeneity among newly
commissioned officers predominated over considerations
of ability. By designing the syllabus to suit these older
public schools with their presumed ability to instil
notions of discipline, consideration and gentility, the
army again sought to ensure that candidates would possess
the appropriate social characteristics.(52) Consistently,
character was preferred to intellect, and in the absence
of any compulsory professional training for these new
officers, nothing was done to ensure their functional
competence.

Even the suggestion that officers should be examined
before they were promoted, proved to be ineffective in
practice. The vaguely-worded regulation was rarely enforced, so that there was little incentive for the officer
to study his profession. There was, moreover, while the
existing system of promotion by purchase continued to persist, no reason why these officers should try to acquire a
high standard of professional attainment. In this respect,
English practice differed considerably from that of Continental countries. In the Sardinian system, all officers
before promotion to the rank of captain were obliged to
attend the Officer's School at Ivres in Piedmont for a
year. In Prussia, all officers attended one of the nine
Divisions-Schule. Only in France, where some one-third
of the regular line officers were drawn from St Cyr, was
there a state of affairs which paralleled that in Britain,
but even here the remainder of the officer corps were
drawn from the ranks of long-serving non-commissioned
officers, qualified by practical experience. It was in
the early Victorian army that the lack of incentive was
most clearly marked, for this was a situation in which the
officer either bought his promotion, or waited for the
seniority rule to run its course, or hoped for a wartime
situation in which either the death of others created
regimental vacancies or his own bravery earned him a
brevet promotion. Here, the army and the Scientific Corps
voiced a common complaint, for in neither group was promotion a question of merit. There was of course no promotion by purchase in either the Artillery or the Engineers,
but there was a comparable lack of incentive because of the
seniority rule which was evidenced by the ages of officers
in the various rank categories. The relatively high mean
age of officers during this period is shown in Table 13,
(53) and it is an interesting comment on the popularly-
held belief that officers, particularly cavalry and Guards
officers, were 'young boys' who had bought promotion at
an impossibly early age.

TABLE 13   Average ages of officers of the scientific corps and army: 1838

	Artillery No.	Age	Engineers No.	Age	Cavalry No.	Age	Guards No.	Age	Line No.	Age
Lieutenant-Colonel	42	54	25	49	29	45	68	42	154	49
Major	-	-	-	-	31	45	-	-	226	43
Captain	80	49	40	45	155	32	86	28	1089	37
Second Captain	90	44	40	42	-	-	-	-	-	-
Lieutenant	167	27	80	28	240	27	57	20	1626	29
Ensign/Cornet/2nd Lieutenant	29	20	19	18	122	21	-	-	899	21

At the same time, it was possible to rationalize this lack of incentive on the grounds that an academic education and training did not necessarily produce those qualities which were demanded of an officer. Despite this criticism made of 'character' as a criterion of ability, in the reports on the Home and Indian Civil Service, it could be argued that 'intellect' could not be accepted as the only basis for promotion in the military. In this respect, the search within the army for 'tone' reflected more general attitudes within civilian society particularly those of professionals in the 'older' professions such as the Law and the Church. But criticism of the intellectual was taken further in the army than in other civilian occupations. Consistently it was argued that the military needed gentlemen in all its officers' ranks, a thesis which denigrated again and again the need for officers to be academically and professionally trained.

This thesis could perhaps be rationalized in the case of junior officers whose professional training in common with civilian professionals was primarily associated with practical experience. Many of their basic tasks needed little intellectual ability or academic training. But even if this argument were valid in the context of the regimental and corps officer, it is questionable whether it was equally applicable to the inner élite of officers who occupied staff and command appointments. On the Continent of Europe, a formal institutionalized training course was recognized everywhere as a fundamental prerequisite for selection for staff and command activities.

For every major European army a national Staff College
had been established, entry to which was by competitive
examination. In all the colleges, the educational and
training programme was extensive in its scope. Everywhere,
the entry of the individual into the inner élite of the
Staff was dependent on the distinctions gained during the
college course. There were differences in the form of
these national colleges, but they were all designed to
give every encouragement to the acquisition by officers of
further knowledge, and to give every inducement 'to call
forth the energies of young officers of talent'.(54)

In Great Britain during the early years of the Victorian
military establishment, it was recognized that 'a College,
or Senior Department, for the Staff is the completion, and
may be termed the strongest encouragement of General Military Education'.(55) This recognition of its value was,
however, no substitute for the persistent failure to ensure that education for the Staff was a prerequisite for
selection to that body. Throughout the years which
followed the establishment of the Staff College at High
Wycombe on 4 May 1799, officers of all ranks opposed the
introduction into the army of any form of compulsory education and training before appointment to the Staff. This
is not to deny that the military recognized the effects of
a grave shortage of trained staff officers, a shortage
which became readily apparent whenever the British Army
was engaged in large-scale operations. In the time of the
Napoleonic Wars, it was said that(56)

When an army goes on service, we are so destitute of
officers qualified to form the Quartermaster-General's
department and an efficient corps of aides-de-camps,
and our officers in general have so little knowledge
of the most essential parts of their profession, that
we are obliged to have recourse to foreigners for
assistance for our operations are constantly liable to
failure in their execution.

Subsequently in the Crimean War, the inadequacies of the
Staff were a frequent cause for complaint, but notwithstanding the criticisms which were made, the attitude of
many senior officers was summed up by the Duke of Cambridge:
'I prefer for the staff to have regimental officers. I am
satisfied that the best officer is your regimental officer. . . .'(57)

The reasons for this attitude were complex and numerous.
In part it stemmed from the wish of commanders to draw
their staff from their own circle of friends and relations,
for this use of patronage helped to secure the homogeneity
of the inner élite of commanders and staff. In addition,
it was derived from the wish of the commander to have at

his disposal a number of appointments which he could dis-
tribute as a form of reward. Concomitantly, this attitude
reflected a common opposition to the 'bookish' officer and
a preference for staff officers who were 'all perfect
gentlemen, extremely intelligent, zealous beyond everything, and most courteous to all'.(58) To a very large
extent, it resulted from the failure of the Staff College
to meet its training obligations. When Le Marchant had
planned the curriculum for the Staff College, he had seen
as its aim the task of producing trained officers who
would staff the Quartermaster-General's Department.(59)
But when officers arrived at the College on the first
courses, it was evident that the primary task of the
College was to instruct these officers in basic educational subjects, to supplement or replace the instruction
which they had received at school. The note books of
Captain (later Lieutenant-General Sir Charles) Brooke who
was one of the twenty-six officers to join the first course
of the Senior Department of the Royal Military College in
1799 show clearly the problem faced by the College. The
notebooks, now housed in the Library of the University
of East Anglia, document Brooke's study of mathematics,
fortifications, military history and elementary tactics.
In mathematics, the course began with a revision of the
basic rules of arithmetic, progressing slowly to a study
of elementary algebra, geometry and trigonometry. The
study of fortifications emphasized the importance to the
officer of an ability to produce detailed, decorative pen
and ink washes. Lectures on military history and tactics
were given in French by General Jarry, and it is evident
from the notebooks that the extent to which the student
profited from these lectures depended to a very considerable degree on his ability to overcome the language problem. Even so, while Le Marchant and his protégé General
Jarry, a French Royalist emigré, were in control of the
College they were able to stress the ultimate aims of the
course while recognizing, at the same time, the need for
some form of remedial education. With their departure and
with the subsequent development of the College these
ultimate aims were forgotten. The College lost direction,
and it appeared the Staff College trained for everything
but war.
     A further reason for the opposition of many tradition-
alists in the Victorian army to the Staff College was the
result of the differences between the attitude of the Army
and the Scientific Corps. In the absence of a separate
Staff College of their own, Artillery and Engineer officers
could not attend the Army Staff College, so they were
appointed to the Staff without having undergone any formal

training. These Scientific Corps officers vigorously
opposed the creation of their own Staff College, arguing
that the superior educational achievements of Artillery
and Engineer officers, and the comprehensive nature of
the initial course at Woolwich made such a college an
unnecessary extravagance.(60) Army officers were not
slow to react to this claim. Conveniently forgetting
that all Scientific Corps officers had gone through a
formal training course, in comparison with the small
percentage of army officers who were Sandhurst graduates,
they rejected the former's claim to an educational superiority. At the same time, regimental officers agreed that
a Staff College was an unwanted luxury, a decision which,
in part, was affected by their reluctance to endorse the
Scientific Corps' suggestion that a college was needed
only for officers of less than 'superior ability'.(61)

In rejecting the validity of the Staff College, army
officers were also affected by the pragmatic realization
that few graduates of the College were, in fact, appointed
to either staff or command posts. Of the 216 officers
who graduated from the College in the twenty years from
1836 to 1854, for example, only fifteen were ever selected
for these élite appointments.(62) Because of this known
pattern of future employment, few regimental officers,
therefore, could be persuaded to apply for a vacancy at
the College. The tendency was for students to be drawn
from among those officers who looked upon the course as
an alternative to an unpleasant and inconvenient overseas
posting. Additionally, officers in command of regiments
refused to let their more able officers go to the College,
arguing that they were indispensable, so that a circular
pattern was set up which did little to improve the image
of the College. In this circle, the situation of a limited
number of poor quality applicants, instruction at an elementary level, non-selection for élite appointments and a
subsequent limited number of poor quality applicants, made
it almost impossible for the Staff College to make a major
contribution to the development of military professionalism.

The whole education and training programme within the
Victorian army in this pre-Crimean period, thus suggested
that little was done to meet the professional needs of
the army. In an era of industrial expansion, when technological development, especially in the fields of communication and transportation, were considerable, senior
attitudes towards professionalism appeared to be summed
up in the neo-feudal preference for 'character' rather
than 'ability'. But the army was not allowed to stagnate
in peace, and a growing external pressure forced the mili-

tary to re-assess the validity of its educational and
training programmes. To a considerable extent, this was
part of a growing middle-class pressure to break into
socially respectable occupations which were barred to
them. In their attempt to break into the official world,
the middle classes, and more particularly their vociferous
spokesmen, directed their attack at the entrance and promotion systems which were in operation. Despite the
theoretical, historical and operational factors in favour
of these established systems, an increasing number of
critics argued that employment in the public service
should be determined by merit rather than by patronage.
Repeatedly, it was argued that the aristocratic principles
of ascription should be replaced by achievement and that
interest should give way to ability.

However, it was the impact of the Crimean War, and more
particularly the reports on the war published in 'The
Times', which encouraged virulent criticism of the military. On the basis of William Howard Russell's reports,
'The Times' began its attack on the 'aristocracy who were
trifling with the safety of the Army in the Crimea':(63)

The noblest army ever sent from these shores has been
sacrificed to the grossest mismanagement. Incompetency,
lethargy, aristocratic hauteur, official indifference,
favour, routine perverseness and stupidity reign, revel
and riot in the Camp before Sebastopol.

The immediate reply to these criticisms was a hurriedly
implemented programme of organizational reforms designed
to rectify a situation in which, hitherto, military administration was shared between seven separate departments
each with its own rules and structure. The subsidiary
effect of these criticisms and of the growing volume of
public concern was the institution of the inevitable committee of inquiry set up after every major British military
disaster. But, however inconclusive these committees may
have been in their findings,(64) the example of the Crimean
War prompted a fresh examination of the problems of professional military education. In a scheme introduced into the
House of Commons on 24 February 1854, Sidney Herbert,
Secretary at War, proposed to introduce professional education for officers after they had entered the army. He
intended to appoint garrison instructors at large stations
at home and in the colonies, and although attendance at
these courses was not to be compulsory, a stringent examination in the subjects of instruction was to be sat by all
officers before they were eligible for promotion to the
ranks of lieutenant and captain. The scheme was approved
by the Commander-in-Chief, General Lord Hardinge, and
Treasury approval was given for the estimated annual cost
of £2,000.

The impulse to reform the military system was by no means abated. In rapid succession, three reports recommended changes in military education.(65) There was ample evidence of a growing public concern with the apparent shortcomings of the pre-Crimean educational system in the military. It was clear that boys who entered Sandhurst or Woolwich at the age of thirteen or fourteen or who purchased a commission when they were no more than fifteen, had only partially completed their education. Even the introduction in 1849 of the examination for purchase candidates had scarcely improved the situation, for prospective officers were frequently sent to cramming institutions which rarely widened the candidates intellectual horizons. At the same time His Royal Highness the Duke of Cambridge became Commander-in-Chief, a post he was to hold until 1895. Despite the criticisms which were later made of him, Cambridge at this time was young, enthusiastic and keen to follow the Continental example of giving Staff and other officers systematic training in peacetime. The amount of support for the Continental model was very apparent in the opinions expressed in official and unofficial sources from 1856 onwards. On the Continent, specialized military training normally began only after the potential officer had completed his general education. To begin this training, candidates had first to pass a competitive entry examination which was based on the national educational system. It was these elements from the Continental model which were incorporated into the various schemes of educational reform introduced in 1857 and 1858. In effect, Sandhurst and Woolwich were to become military colleges for the training of officers rather than schools for the education of schoolboys.

In addition, the Duke of Cambridge laid down new educational qualifications for staff officers and doubled the number of students at the Senior Department of the Royal Military College, Sandhurst which now became a separate Staff College at Camberley. This was the start of almost twenty years of fitful military reform, as increasingly the Victorian military establishment found itself subjected to external pressure. On the one hand, the example of Prussian expansion in Denmark and Italy brought home to the British public the extent to which their army had lost the dominance in Europe which it had formerly enjoyed. At the same time, the lessons of the American Civil War illustrated very clearly the differences between the campaigns of Crimea where Britain made do with the resurrected army of Waterloo and the Peninsula, and campaigns in which mass armies and industrial power generated long wars of attrition. War could no longer be left to the amateur. The

enormous changes in the nature of war brought about by
the inevitability of progress demanded an increased professionalism and this, in turn, necessitated improved
military training and professional education.

5

The Search for Professionalism

One of the more important subsidiary effects of Prussian
successes in the campaigns of 1866 and 1870 was that the
German victory drew attention to the contribution made by
a high level of professionalism to the creation of an
effective and efficient military. The example of the
Prussians in forming the first European army of the industrial era was not lost on the rest of the world. In 1878
the French, for example, followed the Prussian lead by
opening the Ecole Supérieure de Guerre where, under Lewal
the first commandant, a functionally and highly professionally orientated course sought to 'place officers face
to face with the eventualities of war, with the unexpected,
in order to form their judgments'.(1) Even in the USA,
where military tasks were primarily those of maintaining
internal law and order, an intellectual awakening stimu-
lated an inquiry into the needs of a professionalized
corps of officers. In 1877, for instance, after a tour of
Prussian military establishments, Brigadier-General Emery
Upton reported that the existing military academy at West
Point did not afford officers, 'the means of acquiring a
theoretical and practical knowledge of the higher duties
of their profession'.(2) He accordingly proposed the establishment of infantry and cavalry staff schools, backed
by the resources of a war college, to produce a pattern of
training and education which would emulate that of Prussia.
    In Britain, military theorists, senior army officers
and politicians similarly attempted to isolate the essential features of the Prussian system, in the hope that the
introduction of these into the Victorian army would produce
a comparable level of professionalism.(3) The initial
problem which many of these analysts faced, however, was
that they were unable to understand the professionalism
which had ensured the Prussian success. Thus while many
contemporary commentators endorsed the statement by Sir

Harry Verney MP that, 'My days with Moltke left a strong
impression with me of the perfections of his system of
training',(4) few agreed with the later conclusion of
Spencer Wilkinson that the foundation stone of professionalism was an effective training and educational programme.
(5)

   This limited appreciation of the basis of the Prussian
system was partly the result of a military reluctance to
profit from alien experience. Robertson, for example,
recounts an incident in which the Duke of Cambridge, taking
the chair at a lecture given to officers of the Aldershot
garrison on the subject of foreign cavalry, commended the
lecturer to his audience with the words, 'Why should we
want to know anything about foreign cavalry? We have a
better cavalry of our own. I fear, gentlemen, that the
army is in danger of becoming a mere debating society.'(6)
This lack of understanding also followed from a considerable misinterpretation of the tactics and strategy which
contributed to the Prussian success. Thus in Britain, as
in France, the apparent significance of the infantry
assault as a means of ensuring victory, hindered any
general examination of the extent to which the Prussian
victory really resulted from their mastery over problems
of mass organization and movement.(7) To this, there were
some exceptions. One British theorist who rejected the
emphasis placed on the importance of the infantry assault
was Colonel G.F.R. Henderson who argued that 'the Germans
relied on fire, and on fire alone, to beat down the enemy's
resistance; the final charge was a secondary consideration altogether'.(8) But even Henderson did not consider
the wider questions which needed discussion, and he, like
many others of his contemporaries, was too short-sighted.
In Britain, as in France, where Colonel Ardait du Picq's
book 'Etudes sur le combat' had a lasting effect on military thought,(9) most analysts consistently considered the
tactical methods adopted in the Prussian army rather than
look more deeply at the Prussian system. This failure to
recognize the size of contribution to victory made by the
Prussian General Staff was perhaps understandable, since
no contemporary civilian industry or public administration
had faced and overcome problems on this scale, but it did
mean that, with some exceptions, few people were aware of
the real needs of professional military education and
training. Wolseley, as one of the few 'thinking' generals
in the Victorian army did try to explain these needs to
the politicians. He made his point very forcibly to the
Secretary of State for War, Lord Hartington:(10)

   The purchase system gave to officers an undefined and
   unwritten, but a practically recognized, right to

> obtain the command of regiments without reference to
> whether they were or were not fitted for the position.
> That system was abolished at a cost to the nation of
> some millions, and in my opinion, it is but right that
> the nation should obtain in return the reform it ex-
> pected . . . namely, that none but competent and pro-
> perly educated officers should be selected.

Wolseley was an exception to the general rule. Yet the
failure to understand the changes taking place in the
pattern of warfare and in the needs of new forms of armies
was not entirely the fault of the army. The attitudes of
many officers were a reflection of the persistent opposition which existed in Great Britain towards any form of
professional training which seemed to conflict with the
objectives of public school education. These, in terms of
their academic aims, had been summarized by Edward
Thring in 1864:(11)

> Let the mind be educated in one noble subject. If this
> subject also embraces a wide field of knowledge so much
> the better. The universal consent of so many ages has
> found such a subject in the study of Latin and French
> Literature.

These aims differed very considerably from those prevalent
in Germany, where in the 'Realschule', developed out of the
'Hohere Burgerschule' of the early part of the nineteenth
century, a more varied curriculum was designed to meet the
needs of future professionals in the Civil Service, industry and commerce. To a considerable extent these German
educational objectives were derived from a particular interpretation of 'Beruf'. Widely translated as 'profession',
the term carried with it religious overtones which suggested that a wide variety of occupations from surgeon to
chimney sweep were 'callings'. Success in these fields
was dependent on entrants having undergone a preliminary
education and training process which prepared them for this
sense of vocation.

In contrast with this German attitude which stressed
the importance of the relationship between a developed
professionalism and a purposive school curriculum, British
attitudes, evidenced in the reports of the Clarendon
Commission and the 1868 Schools Inquiry Commission under
Lord Taunton, continued to emphasize the significance of
character training as an educational aim. Little was done
to meet the demands of social and technological change,
and the education system as a whole was slow to react to
the professional needs of expanding occupations. This
conclusion was explored in some detail by J.F.D. Maurice,
who, writing in the 1850s, argued that 'We have so much
dreaded to make the Education of our Schools and Univer-

sities professional that we have kept it at a wide,
almost hopeless, distance from professional life'.(12)
The education of the upper classes, therefore, still
seemed to be designed to meet the political and social
needs of a ruling élite. The secondary schools, which
catered for the needs of the middle class could have
adapted to environmental demands, but they either adopted
the curriculum of the old public schools, or degenerated
into a position where they were 'certainly not superior to
the best conducted elementary schools'.(13)

So the objectives of character training continued to
dominate the pattern of English education for the next
fifty years, to the extent that even after the Boer War
there was an over-readiness to accept as irrefutable the
association between the educational aims of the public
school and the development of professionalism. There
was, in the words of a 'Times' leader, which examined
military professionalism, a simple belief in the ideal
fitness of the public school curriculum, and too great
a readiness to be guided by public school masters.(14)
The effect of this attitude in the Victorian army, an
attitude which was epitomized in the evidence given to
the Dufferin Commission by W.J. Cowthorpe, Senior Civil
Service Commissioner, who expressed his belief that 'Our
public schools, as at present constituted, furnish the
very best material for the officering of our army',(15)
was to generate a consistent willingness to accept an
imperfect public school educational process as a satisfactory form of professional training. In the context of
military education, it could be concluded that this was
imperfect because it was 'an intellectual discipline which
practically ignored Continental languages and the study of
English language and history, and under which intellectual
progress was of less account than athletic distinction'.
(16) Moreover, the new grammar schools, established under
the 1902 Education Act, chose to follow the traditional
pattern, so that the effects of the public school tradition were felt not only in the Victorian army but also in
post-Victorian society and its contemporary military.

In addition to the constraints thus placed on the development of a military training programme by the generally
accepted principles of civilian education, the former
programme was affected by a specific opposition to the
very existence of separate military colleges.(17) Initially,
this stemmed from a belief that the cost of maintaining the
colleges, particularly Sandhurst at a net charge of
£17,000 a year, was excessive and an unwarranted expenditure of public funds. In addition, it was argued that
military cadets were generally idle,(18) so that in terms

of cost effectiveness, the financial outlay could not be justified. A more fundamental opposition to the existence of the colleges, moreover, was associated with the belief that these colleges, noticeably Sandhurst in comparison with Woolwich, did nothing which could not be better carried out at one of the major public schools.(19)

The effect of this belief on the external evaluation of military professionalism raised several interesting points. Of the criteria of professionalism, one of the most significant, because it affects the professional status of an occupational group, is the need for the group to convince non-members of the exclusive nature of the task activity. This, in turn, presupposes that acquisition of the skill requires special training, that only a small number of selected people are capable of performing the task, and that a specialist training is needed by new entrants to the occupational group before they can carry out the professional activity. On this basis, many of the developing Victorian professions sought to achieve a licensed monopoly in a given area, so that by 1900 a number of occupations were protected by Statute — Solicitors, Pharmacists (1852), Medical Practitioners (1858), Dentists (1878), Veterinary Surgeons (1881) and Patent Agents (1888). This, in turn, enabled the group to establish training and educational criteria which were used to evaluate the acceptability of potential members and, increasingly, these criteria were associated with the provision of formal training courses in selected institutions of higher education. Against this, if the belief that officers could be trained effectively in a public school was commonly held, then this suggested that a large number of people rejected the claim of the military academies to any form of professional exclusiveness. These critics were rejecting the claim of the military to be recognized as an identifiable professional group, because they neither accepted the need for specialist training nor agreed that this was a prerequisite for the performance of the group's task. Their reliance on the efficacy of a general school education as an effective basis of professionalism considerably weakened the status claims of the military, but while it was perhaps natural that the headmasters of public schools should be biased in their conclusions, it was an interesting comment on contemporary assessments of the values of a military college, that senior officers could also hold a similar point of view, allegedly based on their experience. As Lieutenant-General Lord Paulet told the 1868 Military Education Commission, 'I know that when I commanded a regiment, I would rather have taken an Eton boy, from choice, than a Sandhurst boy, to make an officer.'(20)

   The reasoning underlying this military attitude towards
its own academies was complex. It was the result, in part,
of the argument that the 'moral tone' of the military college was inferior to that of the public school.(21)   It
was also believed that the expense of Sandhurst was un-
justified, because there was no evidence that cadets made
better officers than boys who entered the army directly
from public schools.(22)   From statements made by Major-
General P. Herbert, MP, and other officers of a similar
opinion, to the 1868 Military Education Commission, it
would appear that they thought that education could be
best carried out in the environment of a closed community.
This, they argued, encouraged cohesiveness and morale,
and developed a 'good tone', so producing 'a thorough
English gentleman — Christian, manly and enlightened —
a finer sentiment of human nature than any other country,
I believe, could furnish'.(23)   Sandhurst on the other
hand was considered by them to be defective as a closed
community, because it was 'contaminated' by extra-educa-
tional influences, with the result that the accepted
objectives of education, noticeably the development of
character, could not be readily achieved.(24)   It was
immaterial to this argument that these influences were
predominantly military in character, or that the desired
cohesiveness was that of a military force and not that of
a public school, and the basic thesis was consistently
used as a criticism of the military academies.
   It is now clear that the influence of the public school
tradition which motivated these conclusions, was far
greater in these years between 1860-70 than was justified
by the number of officers who had actually been educated
at these schools. Until the pattern of recruitment changed
in the late Victorian and Edwardian period, relatively few
officers were drawn from the Clarendon schools. This was
made clear in 1864:(25)

   The number of public school boys who enter the army is
   not large. Of 1,976 candidates for direct commissions
   within three years, 122 only had been at any of these
   schools. Of these 102 succeeded and 20 failed . . .
   of 96 who passed at the first examination, 38 came
   immediately from school, 58 had intermediate tuition.
   The public school candidates for Sandhurst during the
   same period, were 23 out of 375; the proportion who
   succeeded being also here much above the average. Of
   18 who succeeded, 11 came straight from school; of
   five who failed, only one.

TABLE 14 Known educational institutions attended by successful army examination candidates during the period 1855-67

Years(s)	Nature of examination	Educational Institutions Attended						Number of successful candidates
		Clarendon Schools	Taunton Schools	Institutions		English Universities	Others	
				Private	Irish and Scottish			
1855-8	RMA	10	35(19)	39	48(31)	8	6	145
1857-8	RMC	1	11 (3)	6	2 (0)	1	8	25
1866-7	RMA	16	76(52)	104	15 (2)	1	37	162
1866-7	RMC	2	73(46)	194	16 (0)	0	49	334
1866-7	Direct Commissions	102	78(36)	497	18 (4)	21	37	621

Sources

I   RMA examination (for August 1855, January 1856, June 1856, January 1858) figures derived from P.P. 1857-8, XXXVII (234).

II RMC examination figures for 1857-8 derived from P.P. 1857-8, XXXVII (195).

III All other figures derived from P.P. 1870, XXIV (25), Appendix XIII.

Notes (see page 140)

TABLE 14 (contd)

Notes

1   Clarendon schools comprise Eton, Harrow, Rugby, Shrews-
    bury, Westminster, Winchester, Charterhouse, St Paul's
    and Merchant Taylors.
2   Taunton schools comprise the endowed grammar schools
    and proprietary schools listed in the 'Report of the
    Schools Inquiry (Taunton) Commission P.P. 1867-8',
    XXVIII, Part i (3966) together with other endowed
    schools not listed in the Report but referred to in
    Gardner's book 'The Public Schools'. The figures in
    parenthesis refer to the number of candidates within
    this category who attended the proprietary schools of
    Cheltenham, Clifton, Brighton College, King's College
    School, and Rossall, and the two endowed schools,
    Marlborough and Wellington College.
3   Private institutions are defined in the Taunton Com-
    mission Report, and include 'cramming' institutions.
4   Irish and Scottish institutions include both schools
    and universities. Figures in parenthesis refer to the
    number of candidates from Trinity College, Dublin.
5   'Others' include home instruction, foreign institutions,
    and those candidates whose origins are unknown.

From Table 14, which is based on that compiled by Ian
Worthington, it is clear that the contribution made by
the Clarendon schools to officer recruitment was far less
than that made by other educational institutions. Even
so, the influence of these schools was paramount. The
influence of the public schools was perhaps less marked
in the Scientific Corps where an even smaller proportion
of public school boys sought admission to the Royal Academy at Woolwich. This was in part a reflection of the
refusal on the part of the public schools to teach science
subjects. In commenting on this, the Public Schools Commission pointed out in 1864 that 'natural science was
practically excluded from the education of the higher
classes in England'.(26) As a result, of the candidates
who were successful in the Woolwich entrance examination
during the years from 1858-67, less than 10 per cent came
from the 'Clarendon' schools. Harrow and Eton each contributed two candidates. Rugby, where in the early 1860s
a tenth of the boys did science, produced one candidate.
No candidates came from Shrewsbury, Westminster or
Winchester.(27)
    Toward the end of the century, however, the public
schools were of increasing importance as a source of
military recruitment. In 1883, for example, 33 per cent
of the entrants to Woolwich and 12 per cent of the entry

to Sandhurst came from the public schools. Between 1896
and 1900, these percentages had increased to 78.7 and 55.4
respectively,(28) the increased contribution of the public
schools to Woolwich reflecting, perhaps, the introduction
of science subjects into the curriculum at schools such
as Harrow, Mill Hill, Clifton, Cheltenham and even Eton.
But irrespective of the precise percentage of potential
officers who came from these fee-paying boarding schools,
the influence of the public school tradition had a persistent effect on the development of military education
and training programmes.

Initially, this influence made it difficult for the
Victorian army to establish the standard of general
education which was needed by aspiring officers as a
basis of their professionalism. Theoretically, the standard was evaluated in the entrance examinations which
candidates sat for admission as a cadet to Woolwich or
Sandhurst, or for direct entry from civilian life. By
1870, the pattern of these examinations was well established. Candidates for Sandhurst, for example, who had
to be between the ages of sixteen and nineteen, were
examined in five subjects. Elementary mathematics and
English were compulsory, and the candidate further
selected three optional subjects from a wide list which
included modern languages, history, natural sciences,
experimental sciences, and geometrical drawing. The
influence of public school curricula was seen in the way
in which papers in these subjects were 'weighted'. A
candidate needed to score 900 marks from the optional
subjects to be successful; however, a total of 3,600
possible marks was allocated to the classics paper, in
comparison with the 1,200 which could be obtained on any
one of the remaining papers.(29) This gave the classics
student a considerable advantage, an advantage which was
endorsed in the identical regulations which existed for
the examination of the direct entrant to the army. Even
though Woolwich demanded that its candidates showed
evidence of a higher standard of proficiency in mathematics, the classics papers were of almost equal importance. The 1868 regulations, for example, laid down that
3,500 marks would be the optimum available in mathematics,
3,000 in classics and 1,000 in each of the other subjects
of the examination. To be successful, a candidate was
required to score 700 marks in mathematics and an additional 1,800 marks from any four other papers. A candidate
who scored half of the classics marks or 1,500, thus required only 300 marks from his other three papers to obtain
a pass mark. A similar weighting was placed on the classics papers which could be sat by candidates for the

examinations for the Civil, Foreign, Indian Civil and
Colonial Services, for in 1870 possible marks for Greek
or Latin studies were twice the totals for French or
German studies or political economy.

In an attempt to investigate the influence of the
public school curriculum with its traditional emphasis
on the study of classics, the 1868 Military Education
Commission asked seven of the staff at the Royal Military
Academy, Woolwich, for 'their opinions regarding the
present regulations for admission to Woolwich'.(30) With
two exceptions, the lengthy replies urged the imposition
of a higher standard of mathematics, whilst arguing that
the marks allocated to the classics papers were too high.
The two exceptions, however, showed the insidious effect
of the classics tradition. Lieutenant-Colonel Milman,
Captain of a Company of Gentleman Cadets, thought that 'it
seemed hard that after entrance, a cadet should lose all
profit from his knowledge of the classics. As it is a
great object to get young men with a good liberal education, I think some arrangement should be made by which
cadets might carry on a portion of the marks gained in
classics at the entrance examination and thus be on a
more equal footing when in the Academy.'(31) A similar
argument was advanced by the Professor of Military History,
Captain H. Brackenbury (1837-1914), who felt that 'Mathematics are carried too far'.(32)

This division of opinion reflected a more general
educational controversy which was of considerable importance in the Victorian period. On the one hand, it was
argued that there was a need for secondary education to
be as general as possible. The wider the base on which
specialization was built, the greater, it was argued,
was the possibility of meeting contemporary demands for
change. This was the 'liberal education' thesis, but the
force of its validity was weakened by the extent to which
'liberal' was interpreted to mean the traditional public
school curriculum. In contrast with this point of view,
it could be argued that an often overlooked task of education was to provide a pre-vocational course, in the
guise of academic education, for future professionals.
The force of this argument was weakened, however, by the
distinction which was drawn within schools between 'training' and 'education', so that 'training' was evaluated as
being in some way inferior to 'education' and thus more
suitable for less able pupils who could not cope with the
intellectual demands of the traditional syllabus.

This conflict of ideas affected industry and commerce
as much as it affected the military establishment, and it
was very evident that a large number of headmasters and

educationalists were not prepared to accept the argument
that the function of a school was to prepare students for
their post-school occupational careers. For the military
this argument had been expressed with a great deal of
force by the Commissioners appointed in 1856 to consider
the form and nature of the training needed by officers of
the Scientific Corps:(33)

> The great schools of the country will perform the same
> service . . . to an academy for young scientific
> officers as they do for places which give a specific
> education for other professions; they will prepare
> for it.

But in rejecting this point of view, the feelings of many
headmasters were similar to those advanced by the
Reverend Dr Temple, Headmaster of Rugby School and subsequently Archbishop of Canterbury. In his evidence to
the 1868 Military Education Commission, he insisted that
the best form of education for potential officers was
precisely the same kind of education as that given to any
other group of boys. In many ways Rugby was an atypical
school, for after 1864, in view of the recommendations of
the Clarendon Commission, all boys studied some science.
(34) Even so, Temple was bitterly opposed to the creation
of a 'modern department' at Rugby in which boys could
study mathematics, physics and languages to prepare them
for entry into the military or other professions. To
Temple, as to other Victorian exponents of the classics
tradition, these subjects were 'soft' options:(35)

> This modern department is exceedingly liable to get
> filled up with a considerable number of stupid boys,
> because these stupid boys do not get on well in their
> Latin and Greek, and then there is a strong temptation
> both to masters and parents to put them over into what
> seems to be, and what is to a certain extent, a more
> easy system.

Coming from the Headmaster of Rugby, these views were of
considerable importance, and they had an exaggerated
influence on discussions about the relevance of school
education to professional training. Fortunately for the
contemporary development of the professions, some schools
at this time, noticeably Cheltenham College and Wellington
College, tried to ensure that part at least of their
curriculum was designed as a preparation for subsequent
professional training. At Cheltenham, for example, the
Modern Department was established for the express purpose
of preparing boys for their future military career. The
teaching of certain subjects, not traditionally part of
the public school curriculum, was deliberately emphasized,
so that the potential officer studied 'Woolwich' mathe-

matics in which more time than usual was spent on the
working of practical examples, as well as the physical
sciences and the basic rules of English grammar. At
Wellington College where, in 1868, 96 of the 312 boys in
the school were preparing to go into the army, the Headmaster, Dr Benson, believed that the purpose of a Modern
Department was to ensure that the young man of seventeen
or eighteen, on leaving school, would be well-grounded in
all the elementary subjects of military education.

Yet despite these advances a number of problems remained. The creation of separate Modern Departments for
the exclusive education of potential officers suggested
that the military, as an occupational group, was different
from the remainder of the parent society. The provision
of this type of education could then be strongly criticized
on the grounds that the initial educational experience of
these officers contributed to their belief that the military differed from all other institutions in society.
Their reaction to external demands for improved military
professionalism could then be associated with the extent
to which their education had hitherto been tailored to
their needs. But any emphasis which was placed on the
need for professionals to follow the general pattern of
education had the effect of reducing considerably the size
of the group from which potential officers could be selected. In theory, the examination was open to all suitable
applicants irrespective of the school they had attended,
but the set subjects favoured the boy who had followed a
classical education, the traditional strength of the public
school. This was a limitation that existed independantly
of the theoretically widened base of recruitment which
might have resulted from the abolition of the purchase
system. The educational barriers thus became as effective as any ascriptive barriers in limiting the number of
entrants into the military. This was particularly so,
since in many cases only those aspirants who could afford
to study at a military crammer were able in any event to
pass any of the three entrance examinations. This was a
point which was constantly stressed by many critics of the
'cramming' tradition:(36)

My general impression is that, as regards a great
many of the public schools, youths who come up for
these examinations go to private tutors for a time
after leaving school.

The part played by the military crammers in preparing
candidates for the army was particularly noticeable in
their contribution to the number of successful candidates
for Sandhurst, though for direct entry commissions, it
was matched, in part, by the numbers of entrants who came

from public schools with a Modern Department. As Table 14
indicates, however, a very small minority of aspirants
came from the grammar schools. Since these were the
schools to which the majority of the middle class sent
their sons, the association between the ruling class and
the military was strengthened by this lack of educational
opportunity for other groups.(37)

Most educationalists were opposed to the cramming system. Examiners in the military entrance examinations
claimed that cramming was detrimental, since although it
enabled candidates to overcome the hurdle of examinations
it also meant that the military often recruited students
with little innate ability. More importantly, cramming
defeated the fundamental purpose of the examination as a
qualitative assessment of potential. Certainly, the comments of these critics seemed to be justified by the subsequent performance of candidates who had attended at a
crammer. In the years from 1864-7, for example, there
were marked variations in the respective standards gained
by candidates at entry and on graduation, as is shown in
Table 15.(38)

TABLE 15 Relationship between place on entry and place
on graduation, Royal Military College, Sandhurst

Place on entry	Education 1864-6	Place on graduation
1	A.C. Steele, Bath	11
2	W.R. Smith, Bath	18
3	Elizabeth College, Guernsey	5
4	King William's College, Isle of Man	32
5	Westminster School; C.R. Rippin	31
6	Revd W. Bodley, St Leonards	19
7	No record; did not graduate	-
8	Eton; Dr Bridgeman, Woolwich	20
9	W.H. Carter, Jersey	24
10	A.C. Steele, Bath	14

There were indications, however, by 1880, that the
pattern of recruitment was beginning to change. From
1876 to 1882, for example, 288 candidates of whom 150
passed sought entry to the Royal Military College from
the universities. This figure was small in comparison

with the 6,000 school candidates for Sandhurst and the
2,743 for Woolwich, but it was the beginning of a trend
which continued to gain momentum. It was evident, too,
that the public school system was in a state of transition as there was a move away from the traditional
emphasis on the importance of a classical education. This
was often thought to result from the influence of the
emerging professional associations who frequently established their own external entrance examinations. The
Reverend Osborn Gordon, Censor of Christchurch Oxford
considered this point in some detail in his evidence to
the 1868 Military Education Commission:(39)

> The examinations for the Army, and other positions
> under government will have considerable influence on
> the general education of the country; and, in fact,
> the number of inquiries which are constantly addressed
> to me are evidence that this influence is already felt.
> The curriculum of studies is gradually being enlarged,
> and subjects which were hardly noticed 20 years ago
> have now gained a recognized position and will become
> more prominent hereafter.

In particular, the army entrance examinations were considered to have had a marked effect in promoting the
demand for mathematics in schools and in increasing the
demand for some form of instruction in the experimental
and natural sciences.

It was thus expected that the effect of these and
similar examinations on the general school education of
the country, would be to 'raise the whole course of instruction of the school . . . scholastic training generally
will eventually improve, in a corresponding degree,
throughout the country'.(40) Indeed there was ample evidence in later years of the establishment of wider based
courses of learning in many schools, as at Wellington
College, where the Modern Department under the direction
of Mr W.H. Eve, a Fellow of Trinity, Cambridge, threatened
to eclipse altogether the Classical Side.(41) Thus in
1895 the Royal Commission on Secondary Education found that
in the Public Schools of ancient foundation, more serious
attention was being paid to the teaching of modern subjects, particularly where this was for the preparation of
candidates for army commissions.(42) At Wellington College,
for example, the Master, Dr Pollock, a future Bishop of
Norwich, had developed further the Modern Department. In
February 1895, he organised a final 'Army Class' in which
for their last eighteen months at the College, pupils were
expected to work harder than their fellows in following a
course of studies designed to give them a comprehensive
pre-professional education. Results in the army entrance

examination improved drastically, the 'crammers' became
unnecessary, and the 'army boys' were able to play a
larger part in the life of Wellington College.

It would be injudicious, however, to conclude that
these changes had an immediate impact on military professionalism. The initial problem which the military was
forced to meet was that a major part of their educational
task was not the development of a planned professional
education, but the more basic and routine function of
rectifying the deficiencies of the English educational
system. This was not limited to the army. In 1906, a
committee set up by the Institute of Civil Engineers to
report on the training and education of engineers found
that preparatory work was weak at the schools in science,
mathematics and languages. Boys going to the Royal
School of Mines from the public schools were similarly
criticized for being inadequately prepared in mathematics
and science subjects. They were, it was said, unable to
use books properly; indeed they were told to read novels
to improve their reading standard, they 'had never learned
how to study'.(43) Thus a large part of the army curriculum had to be devoted to the teaching of those subjects
which had been imperfectly studied by cadets in their
schools or which had been completely ignored. But a more
important problem was that the army itself was not certain
whether a formal professional training was entirely necessary. The force of the 'amateur ideal' continued to be
very pronounced. It was a notion that manners, signifying
virtue, and classical culture, signifying a well-turned
mind, were better credentials for leadership than any
amount of expert practical training. There was a continuing belief that to withdraw a promising student from
his classical studies and to replace these with professional training, was to do his education an injury:(44)

> To anyone who will consider the intellectual exercise
> that a good lesson in first-class Greek or Latin
> authors involved, and who will compare this with the
> formal, though in their way, valuable studies which
> constitute military science, the conclusion would seem
> to be obvious that to drop so high a study as that of
> the classics, just at the time when the mind is begin-
> ning to open to their full appreciation is, education-
> ally, a misfortune.

Statements such as these could be rationalized on the
grounds that the learning of classics was to be advocated
because it developed the mind (the 'faculty theory of
psychology'), but they were also the expression of a more
subtle reasoning. The British political élite, like the
Mandarin Chinese and the Toguwara Japanese, were convinced

that a thorough grounding in the classics was the best
training for a country's administrators, statesmen and
military leaders. This, it could be argued, reflected
the connection between the classics, with their associations of the heroic and imperial spirit of Rome, and the
development of the British Empire. The classical tradition stood for order, authority and discipline.(45)
Essentially, it was an élite or patrician field of study,
since those who had undergone the rigours of the traditional humanist discipline in school and university were
accepted by the majority of their contemporaries as an
authoritative élite.(46)

   The corollary of this point of view was the belief
that practical training was a plebian activity and an
infringement of the status of the gentleman amateur. At
Sandhurst, the belief was reflected, for example, in the
limited amount of instruction given to Gentleman Cadets
in the duties of the ordinary soldier. Unlike West Point
or the Canadian Royal Military College at Kingston where
the cadet, at one time or another during his training,
carried out all the functions of a non-commissioned officer, the Sandhurst cadet was shielded from the tasks
which were carried out by the men whom in future he would
command. His professional education was carried out in
an atmosphere of theoretical rather than practical instruction, with a curriculum which, in stressing the need
to conform to laid-down abstract regulations and rules,
produced a graduate who 'regarded with horror any deviation from a sealed pattern. Little encouragement is given
to originality of mind. . . . The result is that he is
inclined to lose interest in his studies, and to regard
them as a nuisance which need trouble him no more once he
has obtained his commission.'(47)

   The situation at Woolwich, however, was not entirely
similar, for the Royal Military Academy continued to
enjoy the innate advantage of being one of the few institutions in Britain during the late Victorian period
which provided professional courses in engineering and
technical subjects. Engineering courses were provided
in Glasgow, Edinburgh, Dublin, Belfast and in London at
King's and University Colleges and at the School of Mines
and Science. The RMA course, and the standards of attainment upon which it insisted, bore a marked resemblance to
these university courses.(48) Indeed since a greater
emphasis was placed at Woolwich on the practical application of the subjects studied then was possible in many
universities, the Academy course enjoyed an enviable
reputation in the 1870s for its comprehensive character,
a reputation which was confirmed by the distinguished
civil appointments which its graduates filled.

The problems which arose at Woolwich during this period of technological expansion in England, were similar to those encountered by the contemporary military when it is faced with the difficulty of promoting technical subjects at a fairly high level of application. There were, it was argued, two alternative avenues of development at the Royal Military Academy. On the one hand, Woolwich could continue to remain a purely military training school with the courses of study reflecting the particular group needs of the Artillery and Engineers. Conversely, it could widen its objectives and its field of recruitment,(49)

> drawing its students from the whole of the national population, having the mind of the country, the tone of the country and the education of the country as much as possible in them; and then put the Academy as nearly as possible upon the university system.

Either choice had certain advantages to recommend its adoption. The development of the Academy as a purely military school of training would have encouraged the growth of military cohesiveness, and would have met the immediate vocational needs of officers. Conversely, the identification of the Academy with an expanding university system would have encouraged the development of 'true' professionalism, would have minimized any possible skill differential between the military and civilian engineer, and would have more closely integrated military and civilian society.(50) Unfortunately, until a positive choice was made between these two possible patterns of development, the innate incompatibilities of the two courses of action generated professional conflicts and dysfunctional consequences which retarded the full attainment of the potentialities of Woolwich instruction.

This was a conflict between the theoretical and the practical, and between the objects of the primitive and competitive military organization. It led to situations in which it could be said that 'if a man has really high mathematical talent, the sooner he leaves the army and goes elsewhere the better for his own sake'.(51) It led to indecision in determining what course of study and in what depth was required for the training of an officer of the Scientific Corps. There was, moreover, an understandable difference of opinion between the civilian teaching staff at the Academy and their military counterparts as the nature of this course and as the manner in which Woolwich should be organized to meet this suggested new task. To the former, it was readily apparent that the organization of the Academy should break away from its association with school:(52)

The system which commenced when they were mere boys is
still continued now that they enter the Academy as
young men. They are, I think, on average occupied 6½
hours daily in the class rooms, during which time they
are tied to forms and desks like school-boys. If
treated like boys, they will be apt to act as boys,
if as men, they will also act accordingly.

If the Academy were to be developed as an institution
of higher education, then it was suggested that 'the
proposed system of management at the Royal Military Academy should be assimilated more closely to a university
type . . . by giving freer scope to the spontaneous energies of its alumni'. In contrast with this point of view,
the military staff believed that the Academy should be
organized as a distinct military formation:(53)

The organization of a military college should be of the
simplest character, the formation such as to habituate
the cadets to military discipline . . . and to ensure
a watchful interest over their social and moral well-
being.

These were two irreconcilable points of view. The
civilian belief in academic freedom, free expression and
professional control over the course, conflicted with the
military insistence on discipline, subordination of the
cadet and the maintenance of hierarchical control. This
dichotomy was also repeated in the varying opinions held
about the subject-matter of study, but in this context the
differences of preferred opinions did not simply reflect
a civilian-military difference of opinion. It was part
of a much larger national question which contrasted the
needs for practical, vocationally orientated 'schools of
application' with the wish to develop theoretical studies
which had no immediate or directly relevant occupational
significance. As far as Woolwich was concerned, some
military officers agreed with the civilian staff that 'all
classes of mind should have some opportunities for free
development'. Others, however, simply saw Woolwich as an
initial training school attended by cadets prior to their
subsequent commissioning in the Royal Engineers or Royal
Artillery.

As a result of this indecision, and the concomitant
conflict of opinions, developments at Woolwich failed to
reflect advances in civilian professional education.
Increasingly other institutions of higher learning were
established, and the output of trained science and engineering graduates from the universities began to
increase. Just over a third of managers in the British
iron and steel industries by the end of the century, for
example, had some professional qualification, and of these

something over a quarter were engineers.(54)   Whereas in
the 1870s many universities had sought to ignore the
demands of science and engineering, by 1900 the newer
institutions were admitting engineers and were giving
more consideration to science subjects. No longer were
Academy-trained engineers in the favoured position they
had enjoyed earlier in the century when in 1851, for
example, men capable of inspecting classes begun under
grants from the Department of Science and Art were so
scarce that officers of the Royal Engineers and Artillery
were commonly used as the only professionally qualified
inspectors. By the end of the Victorian period, although
Woolwich continually reviewed the nature of its task and
modified its syllabi to meet changing conditions, it had
to be admitted that much of its training was out of date.
(55) The level of military professionalism which was
attained was thus lower than might have been expected
from the innate advantages possessed by the Academy as
the originator of British technical and scientific education. While the standard of academic achievement was
higher than that reached at Sandhurst, to the extent that
of a proposal to amalgamate the two military institutions,
it was said that it would be 'like linking the living with
the dead',(56) unsolved questions of policy limited the
development of a high professional standard. As a result
young officers of the Scientific Corps appeared to give
the impression, when they underwent further courses of
training at the Chatham Royal Engineers Establishment or
at the Woolwich Royal Military Repository that they were
for the most part inadequately educated:(57)

> Young men who have read nothing but what they are
> obliged to read, or can get marks for. . . . Though
> their intelligence is evidently of a very high order,
> and their mathematical attainments beyond my measure,
> the majority do not seem inclined to continue their
> own education from any interest they take in it, from
> a taste for general reading or from professional
> ambition.

These defects did not pass unnoticed. Sir Charles
Dilke, for example, in a series of articles published in
the 'Fortnightly Review' in the winter of 1887 drew the
attention of the British public to the inefficiency of
the military establishment:(58)

> While we have a small efficient white army in India,
> even in India the greater portion of the troops we
> nominally possess are non-efficient, and in England,
> with an equal expenditure, our army in a modern sense
> may be said to be non-existent.

But in striving to improve its efficiency through the
development of professionalism, the Victorian army was
always faced with a number of apparently insurmountable
obstacles. The root cause of many of these was the demand
for financial economy, a tangled morass of claims and
counterclaims in which the military blamed the politicians
for starving the army of money, while the politicians
showed how other European countries got better value for
the amount of money spent on defence. It was a situation
which led Sir Frederick Maurice to comment morosely in a
rejoinder to Dilke's criticisms, 'Our army is, as we
believe, the cheapest in the world for the work it has
to do'.(59) Yet civilian critics of the army could consistently refer to objective evidence, as when they
argued that the Prussians could field nineteen army corps
for a cost of £19,300,000, while for £14,600,000 Britain
could put one or, at best, two corps into the field.(60)
As Dilke commented in the House of Commons in 1893 when
pointing out that Roumania could field four army corps at
a cost of £1,500,000 while Britain spent £17,000,000 on
scarcely three corps, 'We are a great Military Power when
we consider the amount of charge we bear for military
purposes and we are a small Military Power, indeed, when
we consider the result in efficiency.'(61)
    Closely linked with this, however, was the continual
low rate of recruitment, a problem which was particularly
acute by the end of the Victorian period. As the 'Spectator' pointed out, 'It is madness to talk about raising
eight new battalions and not to grapple with the problem
that underlies the whole question, the problem of attract-
ing suitable recruits.'(62) Nor was the problem simply
one of commissioning a sufficient number of officers. The
difficulty in recruiting a rank and file was primarily one
of numbers; for officers it was a more complex problem.
On paper, there was no shortage of aspirants. In the five
years from 1896-1900, for example, a total of 4,245 candidates had sat the examination for entry into Sandhurst
alone, and although some of these candidates were repeating previous attempts, an adequate number of applicants
seemed to be available.
    There were, however, two limiting factors, both of
which complicated the recruiting problem to such an extent
that they had a considerable effect on the wider questions
of military education and professionalism. This relationship was noted by St John Brodrick, the Secretary of State
for War in the House of Commons in 1902, when he argued
that, 'This question of the education of the Army is mixed
up with the difficulty of getting a sufficient number of
officers'.(63) The first of these limiting factors was

that only a limited number of officers could graduate
from the two military colleges each year to fill the
required annual intake of 795 officers. The majority of
recruites to the officer Corps, therefore, had to be drawn
from other sources, as is shown in Table 16.(64)

TABLE 16   Forecast annual supply and demand of army
officers, 1901

Demand		Supply	
Royal Engineers	33	Royal Military College	240
Royal Artillery	100	Royal Military Academy	150
Cavalry	70	Universities	50
Guards and Infantry	460	Militia Regiments	355
West India Regiment	12		
Indian Staff Corps	120		
Total	795		795

   The Royal Military Academy was able, with very few
exceptions, to meet the demands of the Engineers and
Artillery and to supply Scientific Corps officers to the
Indian Staff Corps. The Royal Military College, Sandhurst,
however, primarily met the demands of the Indian Staff
Corps, the Cavalry and Guards, so that the bulk of Infantry
officers, some 375-400, came from the universities or the
Militia Regiments. One effect of this was that less than
half of the intake into the officer corps each year had
gone through a process of professional socialization and
assimilation. Whatever the educational limitations of
Sandhurst and Woolwich may have been, their common
strength was that they created a sense of cohesiveness
which was based on the reference group of the army as a
whole. Militia officers, in contrast, were more local in
their attitudes, for like their predecessors who had
entered the military via direct commissions, their group
experience was exclusively derived from their membership
of a particular regiment. Additionally, these non-
Academy officers lacked even a modicum of professional
training, and this large-scale recruitment of direct
entrants into the Victorian army made it very difficult
to argue that an institutionalised professional training
was an integral part of professional development and increased professionalism.

The second limiting factor was the low educational
standard of many of these would-be officers. The persistent criticism which was made of the professional
standards of the Sandhurst graduate,(65) was, in itself,
an oblique comment on the capabilities of those whom the
Military College had rejected at entry. Yet it was these
failures — 2,697 between 1896 and 1900 — who subsequently
joined the Militia for the express purpose of gaining
entry into the regular army. Since in addition there was
an annual deficit of 500-550 officers against a Militia
establishment of 3,385 officers, the position was reached
whereby, 'The War Office cannot afford to pick and choose
to any great extent, but must take in candidates that
appear suitable, in the opinion of the Commanding Officer
of the regiment, to hold a Militia Commission.'(66)
When faced with this situation, the majority of officers
who commanded Militia Regiments were forced to commission
any applicant who seemed to be fit and desirable irrespective of their potential ability.(67) In theory, these
Militia officers were then obliged to carry out four
months of full time training with their regiment during
the first year of service and a period of one month in
their second year. In practice, however, this attempt to
give the new officer some form of compulsory military
training often failed, for these periods of service were
frequently reduced. In the Bedfordshire Regiment, for
example, it was admitted that officers only attended for
three months in the first year, while in the second year
'there is a great tendency for them to ask for leave off
the training in order to stay with the crammer'.(68)
In addition, it also became clear that much of this
initial training was at a fairly low level of application.
In the Royal Monmouth Royal Engineers, a unique Militia
regiment in that it was the only one of the auxiliary
forces to have associations with the Royal Engineers, it
was admitted by their commanding officer, Lord Raglan,
that 'Recruit officers do 56 days preliminary drill'.(69)
    After he had served 'two annual trainings with the
same Militia', the potential regular officer sat the
Militia Literacy Examination, the subjects of which were
identical with those set for the entrance examinations
into Sandhurst and Woolwich. Opinions differed as to the
difficulty of this Literacy Test. The Headmaster's Conference of 1899 condemned it as an examination: 'Most
practical teachers will agree that the varied assortment
of subjects constitute a test which would be severe for
the most intelligent candidates.'(70)
    In contrast with this tacit assumption that the average
Militia officer could not have been particularly intelli-

gent in view of the large number who failed this test,
a contrary point of view was taken by the Headmaster of
Bedford Grammar School, who thought that it was 'A slight
composite examination in literacy and professional subjects . . . the literary standard required being a low
one'.(71)

   This difference of opinion reflected the traditional
variation of attitudes towards these examinations. The
decision to base the entrance examinations for the military
colleges and for direct commissions primarily on the subjects of general education had practical as well as theoretical foundations. For a start the schools were unable,
and more often unprepared, to provide specialist tuition
in a large number of military-related subjects. An examination based largely on the classics, with allowance made
for several more modern subjects, such as history and
foreign languages, required little in the way of additional
facilities. Understandably, this was more likely to
appeal to public school headmasters, whose primary task
was to prepare candidates for the universities. The Headmasters' Conference thus evaluated the test against the
objectives of the public school curriculum, that is, education in depth in a relatively narrow range of subjects.
Phillips, the Headmaster of Bedford Grammar School, however, was considering the broader issue of occupational
requirements, an attitude which led him to comment forcibly
on the particular needs of the military as a profession:(72)

   The intellectual requirements of an officer in modern
   war seems to be much greater than in the times when
   officers merely led their men with personal gallantry
   and dash. Now, officers have not only to grasp the
   ideas and principles of the past but to modify them
   essentially owing to the changed conditions of the
   present. The intellectual strain on officers in the
   present day, when tactics are so radically changed,
   will be serious, and we ought to see that every cadet
   has such an intellectual equipment as will fit him for
   meeting the difficult problems he will have to solve.
The emphasis which Phillips placed on the intellectual
requirements of a professional thus conflicted with the
often expressed preference for the educated amateur, an
attitude which could be summed up in the comment that
'England, at heart, hates the expert; Germany rejoices
in him'.(73) When faced with these contrasting ideas
therefore, the military establishment was forced to
adopt a compromise, which satisfied neither the tradi-
tionalists nor the progressives. To the former, the
military, in imposing any form of educational selection
for entrants from the militia, was excluding an

unnecessarily large number of suitable applicants; to
the latter, it appeared that the standard sought from
potential officers could justly be said to be 'only a
little higher than the highest standard of the Board
Schools'.(74)

For other reasons, it seemed unlikely that the majority
of officers recruited through the university entrance
scheme would be of any higher intellectual standard.
There were some notable exceptions, among them Douglas
Haig, a graduate of Brasenose College, Oxford, of whom it
was said by General Sir Ian Hamilton that he was not born
into the world with a silver spoon in his mouth, but
possibly with a silver pencil in his hand ready to start
his calculations.(75) But in general, it was alleged that
the majority of university candidates were men who had
failed at Oxford or Cambridge or both. Even if this were
an over-harsh criticism of these officers, the 1899
Regulations did little to encourage the recruitment of
highly-qualified graduates, for commissions were granted
to university students who had passed their first year
examinations. In the unlikely event that there would be
more candidates than there were vacancies, the normal
Sandhurst entrance examination was used as a competitive
selection test, even though the examination was specifically designed to evaluate the potential of the younger
school entrant. If there were no competition, then candidates were simply required to qualify in geometrical
drawing.

From the evidence which was given to the 1901 Military
Education Committee it would appear that many of these
candidates lacked both a general education which would
serve as a basis for the subsequent development of professional education and training, and any noted intellectual
ability. The most which many candidates did at university
was to pass their first year examination. Few partici-
pated in university life, and of candidates who came from
Oxford more specific criticisms were put forward:(76)

As Mods is an easy examination, they generally attend a
course of study under Mr. Craig at Oxford, an Army
tutor, and, as a matter of fact their study is quite
different to that of other undergraduates. They do
most of their work entirely apart from them; their
preparation for Mods is not done by College tutors,
but by Mr. Craig, who is an army tutor, and I think
this is a disadvantage.

These and comparable criticisms of 'the sickening
details of ineptitude, stupidity and short sighted parsimony'(77) which characterized the pattern of Victorian
army education and professional training reflected a

general dissatisfaction with the professional standards
of the later Victorian military. In comparison with the
modern definition of the professional officer as a man
'who takes his work seriously, who studies it from all
aspects, who (above all) has the mind as well as the
aspiration, to think an issue through for himself, from
first to last'(78) the officer of this period was seen to
be a man who gave himself up to sport and amusement. There
was, it was said, little actual work for the officer to
do, for 'there is nothing in the position marked out for
him which can satisfy the natural impulse of a healthy man
towards occupation of some kind'.(79) Lacking incentive,
the officer often developed an antipathy towards any
scheme designed to improve his professionalism, an antipathy reinforced by his previous school experience which
too frequently had stressed the merit of brawn over brain,
and character over intellect. His lack of motivation was
exacerbated by a system which allowed him generous leave
and encouraged his participation in field sports, but even
if he had wished to develop his professionalism, there
were few formal post-graduate courses available, either
of a practical or academic nature. Indeed, as the
Adjutant-General admitted to the Military Education Committee, the existing military system had rendered real
military training absolutely impossible. In his evidence
he was supported by the majority of witnesses who appeared
before the Committee, for they agreed that under normal
circumstances, effective training in regiments at home
could only be carried out in the face of great difficulties:(80)

> The want of efficiency of the young officer in
> professional matters lies in the difficulties which
> the present organisation of the Army throws in the
> way of the effective and practical training of both
> officers and men.

There were several effects of this. In practical
terms, it meant, for example, that the army rarely trained
together as a homogeneous unit, a deficiency which became
all too apparent in wartime when a collection of individual
and often competing regiments were expected to co-operate
as a single body, 'with the inevitable result that in the
opening stages of a campaign, British troops must incur
grave risks of disaster'.(81) In academic terms, it meant
that few Victorian officers attended the Staff College, the
one institution which at this time, as hitherto, provided
post-graduate training for a small number of officers.
This was in many ways surprising, for one lesson which
should have been learnt from the Prussian success in 1870
was that victory for an army in the industrial era

depended not only on individual bravery and valour but
also on the contribution made by the Staff to the creation
of an effective and efficient military organization.

For a number of reasons, however, the lesson was not
learnt. The shortcomings were brought out by the Esher
Committee in 1904:(82)

> While the great powers of Europe were seeking to
> perfect their General Staffs, the arrangements at the
> War Office provided only for the collection of intel-
> ligence, and for the preparation of plans of mobi-
> lization, the branches concerned being passed from
> one high official to another in accordance with the
> views which prevailed at the moment.

The functions of a General Staff were but little understood. To the Duke of Cambridge when he was Commander-
in-Chief, a Chief of Staff was a 'fifth wheel to the
coach', a conservative attitude which led him to object to
Cardwell's proposals to create a Chief of Staff.(83)   But
it was not only the military élite who failed to understand the importance of the part which the staff had
played in the Prussian victories. When the Hartington
Commission of 1888 submitted its reports, Sir Henry
Campbell-Bannerman, a former Secretary of War, refused to
accept the Commission's recommendation that a General
Staff be created. In dissenting from the report, he argued
that England had no use for a staff as other European
countries did. Soldiers with practical experience were, he
argued, far more able to prepare schemes of defence than
any officer or body of officers 'who sit apart and cogitate
upon the subject'.(84)   It was these attitudes which were
reflected through the entire army, so that (85)

> While elsewhere in Europe, central planning staffs
> were being created and command systems were being
> organized to control the sinews of war created by the
> industrial revolution, in Britain the constitutional
> issue continued to preoccupy the thoughts of soldiers
> and politicians alike.

The essential point, which was lost from sight in these
continuous arguments about the merits of civil or military
control over the military administration, was that the
creation of a General Staff also ensured the creation of a
general doctrine of war. This was the crux of the Prussian system, where the General Staff had laid down a
pattern of thought and organization to which every officer
was expected to conform, and to which he would be known to
be conforming. Far from stifling initiative, this promoted independence of thought, for the German example of
1870 had shown that commanders at all levels could be
freed from doubts about their colleagues. The creation

of a General Staff thus created a sense of military homogeneity as well as developing military professionalism,
two desirable objectives which supplemented the more important goal of establishing a general doctrine of war.

The failure to learn from Prussian experience during the
Victorian period not only ensured the absence of a General
Staff modelled on Continental practice.(86) It also meant
that the Staff College, as one of the few institutions providing professional post-graduate education, was deprived
of a sense of purpose. Lacking an objective, therefore, it
was not perhaps surprising that it failed to attract to it
the best among the regimental officers of the army.(87)
Initially this was emphatically denied by the military
élite, but by 1880 it had to be recognized that the area
of choice from which aspirants were selected, was only a
small fraction of the total of all eligible officers.
Statistically, the number of candidates sitting the
qualifying entrance examination fell from an average of
over 40 each year for the period up to 1878, to a total
of 33 in that year, and to 27 in 1879. These figures were
not exceptional. Between 1858 and 1868, only 346 candidates, of whom 188 were successful, sat the examination.
Against an annual intake of 24 students, the number of
candidates in the years 1866, 1867 and 1868 totalled 34,
28 and 23 respectively.

The paucity of candidates thus made it very difficult
to use the qualifying examination as a means of evaluating
potential ability, or as a method of selecting a limited
number of high-quality applicants from a wide field.
Continually, it was necessary to fill up vacancies,
although toward the end of the Victorian military establishment the number of candidates did increase. In 1895,
for example, there were 59 candidates for 32 vacancies,
but even here the candidates were not a representative
cross-section of the army as a whole. Cavalry officers
were most reluctant to attend the College, although of all
the interest groups in the army, the cavalry, with its
enhanced prospects of filling élite appointments in command or on the Staff, could have profited most from this
post-graduate course. Allenby, for example, was the only
cavalry officer to enter the Staff College by competition
in January 1896, and the first officer from his regiment,
the 6th Inniskilling Dragoons, who had ever done so. A
few officers, such as Robertson, serving in 1895 in India
with the 3rd Dragoon Guards, realized that without graduating from the Staff College future professional advancement was doubtful, but these were the exceptions.(88) For
the majority, cavalry and others, 'The Staff College was
not considered very seriously by the average young officer

or his commander; having passed the course there did not
compare, for instance, with a large private income or an
expertise at polo, both of which were useful assets to a
good regiment'.(89)

The effect of this attitude on the number of candidates
had been noted by the 1880 Committee of Inquiry into the
Staff College which was worried that the College had
failed to attract all the students that it was desirable
to attract, and had admitted a minority of students of
doubtful ability. A very prevalent conclusion drawn by a
number of senior officers was that candidates, because of
the opposition of most officers to post-graduate study,
were 'officers more desirous of escaping regimental duty
and disagreeable foreign stations than of qualifying for
Staff Service'.(90) This was an interesting conclusion,
for it demonstrated group condemnation of the apparently
deviant officer who sought to break away from the restrictive environment of regimental life by attributing to him
motives of self-interest. But for as long as this remained a general attitude, it proved almost impossible to
increase the number of suitable candidates, since few
young officers were prepared to risk the condemnation of
their peers.

At the same time, this condemnatory attitude was not
entirely unfounded, and its adoption could be rationalized.
Many officers who went to the Staff College did come from
regiments on foreign service; some had already been
serving on the staff before they sought to go to the College. For the latter, in particular, it could be argued
that while their experience might have prompted them to
seek further formal training, the fact that they had
carried out their task successfully without training,
suggested there was no justification for a Staff College
course to train them to carry out the job they had just
left. Additionally, since the system assumed that officers of the Artillery and Engineers were sufficiently
fitted for Staff appointments, on the basis of experience,
ability and previous knowledge, without their attending
Staff College, it was very difficult to justify to a conservative colonel the need for his army officers to attend
the College. This assumption moreover automatically con-
tradicted any assertion that Staff College training was a
necessary prerequisite for Staff employment. It reduced
the observed level of professionalism by creating the
impression that such training was superfluous and not
necessary for task performance. It suggested that there
was no reason to suggest that the attitude and behavioural
pattern of the potential Staff officer could only be
changed through a formal training course. It also limited

the status of the College as a role formative institution, since apparently successful role performance was not dependent on the effect of, or participation in, the training system.

When the characteristics of this system are considered, it is not readily apparent whether the effects of the selection procedure and the concomitant low number of applicants dictated the form of the training system, or whether the imperfections of the latter discouraged any larger number of candidates. An added complication which makes it more difficult to reach any conclusion was that many of the faults in the system were the result of its relationship to the more general pattern of contemporary English education. Insistence, here, on the benefits of a broad-based curriculum which perpetuated 'that vast desert of Englishmen "Gentlemanideal", that attidue which prefers amateurs and gentlemen to professionals and cads', (91) minimized the need to develop a professional skill. So the subjects taught at the Staff College seemed to reflect a wish to create as in the Civil Service, 'a professional class brought up in the first place, with fine and governing qualities, but without the idea of science'.(92)

The Staff College curriculum at the time of Cardwell's reforms, for example, was derived from a very broad general programme of education. Studies such as geology, chemistry and physics which were, in theory, optional, became compulsory if the requisite total of passing-out marks were to be obtained. Critics of the system were quick to point out that this system tended to make mere smatterers, jacks of all trades and masters of none (except by accident), but few were able to suggest acceptable alternatives. On the contrary, some would-be reformers wanted to widen the curriculum even further. Colonel Fielding, who was the Assistant Adjutant-General at Dublin at this time suggested a very general course:(93)

> The course should embrace every possible subject of value to the staff officer: every language, likely to be of use, geology, chemistry, electricity, telegraphy, mechanics (in the highest sense), astronomy, practical carpentry and turning, the keeping of accounts by double book entry, in addition to those subjects which more obviously bear on the career of a military Staff Officer.

Concomitantly, it was argued that any adopted training system should be one which encouraged and permitted a student to follow his 'special bent of study', an argument which if it were adopted further ensured the provision of a very widely-based curriculum.

Because of this pressure, and because of the failure of
the military to specify the duties of a Staff officer, the
Staff College curriculum sought to provide a wide variety
of subjects which were ultimately tested by a very academic question paper, in which, for example, the potential
Staff officer was asked to describe 'the general character
of the flora of the Carboniferous period'. This meant the
exclusion from study of more 'professional' subjects such
as military administration, communications and transport,
or strategy, an omission which led Lord Strathnairn to tell
the Military Education Commission that 'The time mis-spent
in sciences would have been far better employed in the
acquirement of the "sine qua non" of war-strategical knowledge . . . I have found all the officers coming from the
Staff College, of whom I have had experience, deficient in
this respect'.(94)

Nevertheless, the Military Education Commission
rationalized the reasons for the teaching of these
apparently non-military subjects by stressing the advantages of a broadly-based curriculum. In accepting the
need for the study of geology, for example, the Council
put forward a very optimistic argument:(95)

It is to be hoped that the officers who are about to
leave the College will avail themselves of the opportunities afforded by their profession, of following up
the subject practically in the field, and of adding
to the present stock of knowledge by valuable observations of their own.

There was some justification for the adoption of this
attitude, since officers were frequently employed in
situations where their non-military academic knowledge was
of use to them. Kitchener, for example, when engaged in
the first modern survey of Palestine, also produced the
earliest maps and site descriptions of archaeological
remains in the area. It can, however, be argued that the
wish to establish a broad base for military educational
programmes was derived from latent anti-militarist feelings
which sought to retard the development of the military professional.(96) In the USA, for example, where a similar
problem, accentuated by the adoption of an isolationist
policy, also had to be faced by the military at this
period, the objectives of military education were not
rationalized until the national attitude toward the officer
corps changed after the Spanish-American War.(97) If the
British failure to improve the professional training of its
military élite were associated with similar latent anti-
militarist feeling, then it can be seen that there was
very little chance of any large-scale improvements in this
area. In the absence of any specific evidence relative to

this point, it can only be noted that successive attempts by individual theorists, such as Lieutenant-Colonel Hamley, Colonel F. Maurice, Colonel G.G. Henderson and Colonel J.T. Hildyard to modify the curriculum and improve the standard of professional education, met with little success. Similarly, successive changes in 1881 and 1886 only re-emphasized the problems faced by the Victorian military establishment in trying to determine the precise nature to the skills which were needed by its élite. In short, it is uncertain how far the observed failings of the Staff College graduates could be attributed to the faults of the system, constrained as it was by the nature of the parent society, and how far it resulted from individual short-comings, in that candidates selected from a small number of applicants came into the College with previous know-ledge, experience, ability and attitudes which could not be modified by the College at this time.

In this context, it is clear that individual motivation was certainly affected by the limitations which were im-posed on post-training employments, for this affected very considerably attitudes towards the value of a post-graduate course of this type. Indeed it was argued that it was the basic lack of incentive which reduced the number of officers who sought entry to the College, and that other problems faced by the College followed on from this failure to attract the best officers:(98)

> I do not think that the results of the Staff College have been altogether so satisfactory as was expected. This effect may be possibly caused from the fact that the best officers in the army do not enter the College: the number of staff appointments are comparatively so few that they see no certainty of being employed after they have qualified.

Until the creation of a General Staff, this conclusion seems to be confirmed by the appointments which were made in practice. A return, for example, of the number of officers below the rank of Colonel appointed to the Staff in the years before 1870 (see Table 17), shows that the greater number by far were not Staff College graduates.(99)

This pattern of appointment to the Staff was continued after 1870, and its effect on evaluation of the advantage of going to Staff College was exacerbated by the apparent exclusiveness of the 'Wolseley Ring'. The existence of this élitist group, otherwise known as the 'Ashanti Ring' or the 'Garnet Ring', suggested that the army could be divided into the 'Ins' and 'Outs', and that membership of the latter precluded selection for appointment to the Staff, irrespective of individual merit. In contrast, membership of the 'In' group, whose origin can be traced

TABLE 17   Number of officers below the rank of Colonel appointed to the Staff before 1870

	Lt.-Col.		Major		Captain		Lieutenant		Ensign		Total	
	SC	NSC	SN	NSC	SN	NSC	SC	NSC	SC	NSC	SC	NSC
1864	1	5	3	6	11	21	1	14	–	16	16	62
1865	–	13	2	12	6	30	–	14	–	5	8	74
1866	–	5	2	5	6	17	2	11	–	4	10	42
1867	–	3	–	–	4	4	–	2	–	–	4	9
Total	1	26	7	23	27	72	3	41	–	25	38	187

SC = Passed Staff College
NSC = Not passed Staff College

Second, it was clear that for a number of reasons, the lack
of post-training experience, the small number of vacancies
at the Staff College — thirty a year — and the restrictions placed on entries from regiments, an insufficient
number of trained and experienced Staff Officers was available. These, as Roberts pointed out 'cannot be improvised',
an opinion which was confirmed by Kitchener who complained
that there was no reserve of qualified Staff officers to
fill vacancies. Finally, it was apparent that the sole
product of a defective training system had been the 'bad-
all-rounder' who lacked the specific specialist knowledge
expected from a professional Staff officer.

However, blame could not be attributed solely to the
Staff officer and to the defects of a training system
which seemed to have done everything except train for war.
Officers and non-commissioned officers alike came under
heavy criticism for their lack of professional ability. Of
the officers, Kitchener noted, 'There appears to be too
often a want of serious study of their profession by
officers who are, I think, inclined to deal lightly with
military questions of moment.'(107) This was an opinion
which was completely confirmed by the subsequent Report
of the Committee appointed to consider the Education and
Training of the Officers of the Army (the Akers-Douglas
Committee) in 1902. In this report, both Sandhurst and
Woolwich came in for criticism. Commenting on Sandhurst,
the Committee noted, 'The cadets cannot be expected to
derive much benefit from their instruction . . . when it
is clearly established that they have absolutely no incentive to work.'(108)

A corollary of this attitude of mind was that many
officers failed to take their profession seriously, and the
effect of this was accentuated by an examination system in
which while the passing-out standard was low enough,
'there is too much reason to fear that even those cadets
who fail to attain this standard have been commissioned
none the less'.(109) Moreover, once the cadet had been
commissioned into his regiment, there was, it appeared,
very little evidence that the young officer took steps to
improve his professional knowledge:(110)

> By no part of the evidence laid before them have the
> Committee been more impressed than by that which shows
> in the clearest manner the prevalence among the junior
> commissioned ranks of a lack of professional knowledge
> and skill, and of any wish to study the science and
> master the art of their profession.

Character, it appeared, was still preferred to intellect, and brawn to brain. One of the few senior officers
to take a contrary point of view was Roberts who, in his

evidence before the Royal Commission into the conduct of
the War in South Africa, argued that 'Brains are even
more important in war than numbers, and in an army which
may contain a large proportion of men who are not soldiers
by profession, trained leaders are especially important'.
(111) But, while the development of tactics showed that
the officer had a completely new role to play, since it
was no longer sufficient for him to bring his men to the
point of battle and then lead them blindly against the
enemy, training and education programmes were still designed to meet the needs of an outdated army which was
little more than a heterogeneous collection of regiments.
War was still looked upon as a game to be enjoyed by
amateur players. Professionalism was largely at a
discount.

    In seeking an explanation for the lower standard of
professionalism at the end of the century, the Akers-
Douglas Committee moved away, however, from a simple consideration of the effects of the training and education
programme. These issues were not in contention. Innumerable reports had previously drawn the attention of the
public to the way in which an imperfect educational process
had failed to produce the expected standard of professional expertise. The concomitant effects of this process
on factors such as the ideology and attitudes of officers
had also been examined in depth. There was ample evidence
that attempts to produce a fundamental doctrine of war had
produced a rigidity of mind which was incapable of responding quickly to changing circumstances. The Boer War
had brought out clearly the effect of a training system in
which the conformity produced in the military colleges was
bureaucratic rather than professional. The enquiring
mind of the true professional was noticeably absent, and
there was no evidence that officers had been 'specially
trained to deal with large questions of preparation for
war and to determine how, most efficiently and economically, to meet the complicated requirements of the
Empire'.(112) The scope of instruction at the three
military colleges had focussed on the details of military
life. Little consideration had been paid to broader considerations at national or international level, so that
there was no indication of the penetration of Staff and
command functions by the whole range of industrial, social
and political questions which concerned the parent society.
What was now needed by the Committee were additional
reasons which would explain away the widespread dissatisfaction with the present state of education, both military
and general, among the officers of the army as a whole.
One explanation which was put forward led the Committee to

believe that failure was derived from a policy of recruitment which, in the Victorian military establishment, as
for the two previous centuries, recruited officers from
a leisured class to whom professionalism was often equated
with vulgar careerism. This was no new explanation. The
barriers against open recruiting to a military establishment in which 'I am sorry to say that the officer wanted
by the Army is only one who can command from £150 to £1500
a year',(113) had been a frequent target for criticism
throughout the nineteenth century. It was a popular complaint. It was an attack on a system in which the 'poor
but meritorious' were excluded from membership of the
officer corps. Yet in many respects it was in 1900, as
in 1870 during the discussions on the abolition of purchase, too facile an explanation of a low standard of
professionalism. Evidence from the example of other professions to which recruitment was more open, suggested
that they too were faced with similar attitudes on the
part of group members who were reluctant to undertake professional studies in depth. If there were a common
denominator, then the basic fault could perhaps be attributed to the initial imperfections of a national educational system which continually rejected any studies
which appeared to be vocational in character.

But in the Victorian military establishment a number of
other factors were additional and more important reasons
for the existence of a low standard of professionalism.
Some of these were readily acknowledged — the lack of
money to introduce innovation or the ineptitude of badly
trained senior generals. Other reasons were however less
readily apparent. In the absence of a Junker caste within
the Victorian military establishment, deficiencies in professionalism which resulted from imperfect training could
not be blamed entirely on the military élite. The size,
organization and efficiency of the army were the responsibility of civil society. The latter was largely responsible
for the enforced professionalism of the army, and it can
be argued that it failed in its duties in many respects. In
two areas, in particular, the civil power could be directly
blamed for the low level of military professionalism. In
the first place, for the major part of the nineteenth
century no clearly defined objectives were established for
the military. This had two effects. On the one hand, this
ommission affected the creation of a viable training and
education programme, for, in the absence of laid-down
goals, such a programme was bound to lack direction. On
the other hand, this omission had a more direct effect on
the standard of professionalism, for one of the characteristics of a professionalized group, was the way in which

members had a well defined area of communal action. The
second area of omission is closely linked with this, for
a high level of professionalism in the Victorian period,
as today, was dependent on the extent to which group
members within this well-defined area of action possessed
a complete monopoly of occupational activity. No such
monopoly existed during this period, and it is apparent
that for the major part of its existence the low standard
of attained military professionalism in the Victorian army
was partly attributable to the way in which the civil
power failed to ensure that the army it controlled mea-
sured up to these two universal characteristics of professionalism.

6

The Task of the Army

Although the British Army was almost continuously engaged
in some form of military operations during the Victorian
period, two areas of uncertainty affected the development
of its professionalism. In the first of these areas there
was, until the preparation by the Hon. Edward Stanhope,
Secretary of State for War from 1887-92, of the Stanhope
Memorandum, no precise definition of the purposes for
which the military existed. Wolseley brought this out in
1887 in his evidence to yet another Royal Commission:(1)

> We have had nothing decided by this country as to what
> the country wants, or as to what our military policy,
> its aims and requirements are. We have a certain
> number of horses and regiments, and those numbers vary
> according to the political exigencies as to what are
> the military requirements of the Empire; how many
> troops we require to have in England, how many we re-
> quire in our colonies, how many in India. . . .

This affected the army in a number of ways. Far too
frequently, when faced by new tasks imposed upon it by the
government of the day, the military was forced to improvise
and 'muddle through'. Military training programmes, which
would, in theory, have prepared for these eventualities,
lacked direction, 'planning for everything except war',
because the end objective was uncertain. In specific
military operations, the army was constrained by the
actions of politicians who were unsure of the limits which
should be set upon the activities of the army which they
had involved in yet another campaign. When seeking money
to expand the army to meet the demands thrust upon it, or
to introduce new weapons, the military was confronted by
the opposition of politicians and public who, uncertain of
the purposes for which their army existed, saw no reason
to waste money on its expansion.

171

At the same time, the force of lay reaction to the
requests of the army was accentuated by the effects of a
second area of uncertainty. Here, the military was the
victim of traditional British opposition to the existence
in peacetime of a standing army. A powerful pressure
group continually argued that there was nothing done by
the regular army which could not be carried out more
efficiently — and more cheaply — by the auxiliary forces
of militia, yeomanry and volunteers, supplemented when
necessary by the Indian Army. The arguments put forward
by this pressure group, particularly those of Sir Charles
Dilke and Spenser Wilkinson in their joint publication
'Imperial Defence' (1897), were very persuasive. They
appealed to the latent military aspirations of many
civilians, and when conscription appeared at one time to
be the only practical solution to the very real shortage
of army·recruits, the volunteer system seemed to be a
cheap alternative.(2) The belief that a standing army was
unnecessary, or that it should be severely limited in size,
increased the general air of uncertainty, making it difficult for the military élite to convince the civil power
of the need to define precisely the purposes for which
the army existed. There were, indeed, individual politicians who considered that the preparation of tactical and
strategic plans, in advance of a political crisis, was
symptomatic of militarism, an ethos which they argued was
alien to the British way of life. This ideological
attitude was, in addition, attractive to a small but
vociferous section of society, because its preference for
an amateur and civilian military had the merit that
acceptance of the arguments which were put forward would
limit the power of the army, and would reduce the privileges of an apparently élitist, self-perpetuating caste.
    These areas of uncertainty, severally and jointly, had
a pronounced effect on the development of military professionalism. This was particularly noticeable when the
army was compared with other occupational groups whose
claim to professionalism already depended for its validity
on the monopoly which they had achieved in 'a well defined
area of communal action'. Not only was it difficult for
the Victorian army to substantiate its claim to a monopoly
of activity in view of the counter-claims put forward by
the pressure group which supported volunteer and auxiliary
forces, but the area of action was also uncertain. In
the absence during the Victorian period of any precise
definition of the purposes for which the army existed,
a number of different and sometimes conflicting interpretations of military goals had emerged. The first of these
was a relatively simple evaluation of aims and objectives

which, supported by both laymen and officers, sought to
simplify as much as possible the uncertainties in this
area. It was based on an interpretation which made no
distinction between political and military objectives,
choosing to forget or ignore this basic principle of
British civil-military relationships. This distinction
was brought out clearly by Hamley:(3)

> War is a political instrument, resorted to by govern-
> ments when peaceful methods have failed. The ultimate
> object of every war, therefore, is political and is
> decided by governments and not by generals who are
> merely the servants of their government.

Instead, this popular assessment, unable to differen-
tiate between strategic and tactical objectives, accepted
without further question the self-evident truth that the
primary goal of the military was invariably the complete
defeat of the enemy's forces in the field, either by
annihilation or by attrition. It was a rational interpretation of the purpose for which the army existed,
provided that wider issues of policy were ignored. The
sole military goal of importance could, indeed, be
'victory', but the dismissal of other considerations
encouraged a very singular evaluation of military professionalism. This to many critics was based simply and
confidently on the presence or absence of ultimate
victory. When its army was successful, then the Victorian
public could assume that the professionalism of its army
was not in doubt. There was little point, for example, in
looking more critically at the cost of the victory in terms
of casualties; in any event, it could be argued that
these had not resulted from professional incompetence but
had been inflicted by a tenacious enemy who had at length
been overcome by a professional and skilled military force.
Losses were to be regretted, but there was ample evidence
in the campaigns of Napoleon, Wellington, Moltke, Lee and
Grant to support the contention of the strategist that
casualties were, in themselves, no reflection on the skill
of the commanding general:(4)

> Such men, whom they [the public] do not hesitate to
> hold up as models, never quailed before any loss, nor
> did they win their battles without incurring enormous
> casualties. It is not our own losses we should count
> but the enemy's.

But in many of the campaigns fought by the Victorian
military establishment even this consideration was irrelevant, since, in a large number of the most convincing
victories, success had been achieved with little loss of
life. As Steevens, one of the most jingoistic of Victorian correspondents, wrote in his account of the Battle of
Omdurman (1898):(5)

By the side of the immense slaughter of dervishes, the
tale of our casualties is so small as to be almost
ridiculous . . . Putting it at its highest, however,
the victory was even more incredibly cheap than the
Atbara.

In a situation such as this, where a vastly outnumbered
British Army had scored a signal success, there were very
few members of the public who were prepared to be more
searching in their evaluation of the victory which had been
achieved. This was a natural reaction. There was no valid
reason why the expertise of the army in this type of operation should have been subjected to criticism. There had
been 'splendid battles', and a public which had chosen not
to enter the military, could readily identify itself with
the actions of its army. This transferred prestige, moreover, was encouraged by the publication of very graphic
accounts of the actions which had been fought. Flushed
with success, very few readers could fail to be inspired
by accounts such as Churchill's vivid description of the
charge made by the 21st Lancers at Omdurman:(6)

The riflemen, fighting bravely to the last were swept
head over heels into the 'Khar' and jumping down with
them at full gallop and in the closest order, the
British struck the fierce brigade with one long furious
shout.

The rider which Churchill added could be overlooked: 'In
120 seconds, five officers, 65 men and 119 horses out of
less than 400 were killed or wounded'.(7) Equally, few
members of the public in the United Kingdom were anxious
to ask why these casualties had occurred at all.   'The
populace has glorified the charge of the 21st for its
indisputable heroism; the War Office will hardly be able
to condemn it for its equally indisputable folly.'(8)
Steevens, among others, provided the answer, but his
analysis of the situation was neither attractive nor
relevant once victory had been assured:(9)

But for the rash handling of the 21st Lancers, the
mistake of putting the British infantry behind a zariba
instead of a trench, and the curious perversity which
sent the slow camel corps out into the open with the
Egyptian cavalry, the losses would have been more
insignificant still.

In this particular incident the one major set-back
which occurred — the charge of the 21st Lancers — was
completely unnecessary, and while it may have had some
defenders in 1898 it has very few today. Indeed it has
been said that it could be coupled with the Balaclava
charge and recalled as 'magnificent but not war'.(10)
But these are criticisms which can more easily be made in

retrospect, and, for the Victorian public, Omdurman, in
common with a large number of other minor campaigns, was
a splendid victory in which an efficient and effective
army had readily defeated its more numerous enemy.

When, however the army failed to achieve a victory, the
public was very ready in its popular assessment of the
catastrophe to criticize the military for its lack of
professionalism. In this situation the question of loss
was a very important one. Now the public argued that
heavy casualties in an army displayed a want of skill in
its commander or in his subordinates. As Hamley commented,(11)

> When — with due allowance for particular circumstances
> and difficulties and after full and fair trial — their
> deeds fail to justify their casualty lists, then it is
> time enough to call them unskillful, not because they
> have lost men but because they have lost them in vain.

It was not only the losses sustained in the Crimea
which encouraged an evaluation of military expertise,
although since the public was really aware for the first
time of what was actually happening, this particular war
had a profound effect on national reaction. Many of the
minor campaigns fought on the North West Frontier of India
also did little to enhance the prestige of the Victorian
military, and, of these, the First Afghan War of 1838-42
was the most infamous example. While this campaign, in
the absence of the war correspondent, did not have as
immediate an impact on the public as had subsequent disasters, the complete loss of the expeditionary force to
Kabul, with the single exception of Dr Brydon, evoked
considerable criticism of the army's expertise. With the
rapid dissemination of news in the later part of the century, however, public reaction became even more marked,
so that a tactical defeat such as that suffered by Colley
at Majuba Hill in 1881 assumed the proportions of a
national disaster. Once again the professionalism of the
army was brought into question and in the spate of criticism which followed this unexpected military reversal,
Colley's defeat was variously explained away as the result
of his incompetence, the demoralization of British troops,
their lack of training in individual initiative, their
inadequate practice in marksmanship, their inability to
make the best use of the terrain, and the crucial weakness
in mounted men.(12) Had Colley's boldness, however, in
taking on a superior Boer force been successful, then there
is little doubt that these shortcomings would have been
overlooked. Certainly, few critics took into consideration the political and logistic constraints under which
Colley fought, though the latter was made particularly

clear when it was subsequently decided to reinforce the
British troops involved with a minimum of six additional
regiments.(13)

Simple evaluations of this kind were commonplace
throughout the Victorian period. In victory, errors could
be overlooked. Little attention was paid to what had
happened to Sir Herbert Stewart in 1885 when, on a night
march from the Abu Klea Wells to the Nile, he lost his
way so that his troops were forced to fight a battle with
their square formations badly disorganized by dense scrub.
No criticism was made of the débâcle in the Ashanti
Campaign of 1873-4, when Butler's independent column lost
all its 1,400 Akim volunteers after an allied force
attacked them in error. Gough's casualties in the First
Sikh War were discounted, although Sir Hope Grant stated
that 'never perhaps in the annals of Indian Warfare has a
British army on so large a scale been nearer to defeat
which could have involved annihilation'. In defeat, on
the other hand, all errors became momentous. General Sir
Duncan Cameron was rightly criticized for his ineptitude
in ordering three successive frontal assaults, which led
to nothing but a futile loss of life, against the Maori
stronghold of Rangiriri, although it may be wondered what
public reaction would have been had the last assault succeeded.(14) Equally, the splitting up of an army into
independent columns was always something of a gamble.
When it succeeded, as at Kumasi in Wolseley's Ashanti
Campaign, it was justified. When it failed, in the way
in which the Boers consistently exploited such a division
of forces with devastating results, then it was considered
to be another example of gross professional ineptitude.

In contrast with this relatively simple interpretation
of strategic and tactical objectives, the assessment of
wider military goals which reflected political considerations was much more complex. This assessment suggested
that the primary role of the military was that of a
'guardian force'. In an embryonic form, some of the
characteristics which could be attributed to the army in
this wider interpretation of the purposes for which the
military existed, resembled those subsequently described
by Janowitz as the 'constabulary' concept.(15) The latter,
which is very pertinent in any examination of the role of
the New Military in the twentieth century, thus provides
a continuity with the experience and traditions of the
Victorian army.

This is clearly seen in the extent to which a major
role characteristic in both instances was, and is, the
employment of officers in politico-military affairs. For
the Victorian army, this was, as Lord Elton points out,

'An age in which, it almost seemed, any stray detachment
of the British Army could be relied on, should occasion
demand, and almost as a matter of routine, to produce a
junior capable of pacifying a frontier, quelling a rebellion or improvising and administering an empire.'(16)
These were young officers like Charles Townshend, who, as
a captain in 1893-5, was the commander of the fort at
Gupis in the extreme north-west corner of Kashmir, close
to the junction of the Chinese, Russian and Afghanistan
frontiers. Here he combined his military responsibilities
with the appointment of District Officer, holding court to
pronounce judgment on the disputes among his native subjects. In situations such as these, the role of the
Victorian officer was very similar to that of his modern-
day counterpart who can find himself, particularly when
engaged on a peace-keeping operation, involved in non-
military activities as varied as making judicial decisions,
acting as a technologist or serving as a civil administrator.

Nor is the similarity limited to this area alone, for
both concepts endorse the doctrine of the minimum use of
force by the military in politically sensitive situations.
This was a doctrine which from time to time, as in the
First South African War of 1880-1, made it very difficult
for the Victorian army to achieve the immediate goal of
defeating the enemy. The politician's wish to use the
smallest number of troops possible in a campaign often
thus conflicted with the military appreciation of the
appointed commander. Thus in the Expedition to Abyssinia
(1867-8), Sir Robert Napier, the local Commander-in-Chief
designate, advised that no less than 12,000 combat troops
would be required, though London preferred a flying column
of a couple of thousand men. Eventually 13,088 combatant
troops were employed, although a total of over 62,000 men
were needed to mount the operation.(17) Similarly, in
the Ashanti Campaign (1873-4), Wolseley's instructions did
not allow him to commit the Government to the use of white
troops. After trying to settle the issue with native
levies to avoid extending the war, he was forced to
advise the Colonial Secretary that the 'very best officers
and the most highly disciplined troops are alone capable
of bringing the war to a successful and speedy issue'.(18)

Equally the experience of the contemporary military in
such operations as those carried out in Northern Ireland
after 1969 illustrates the continuity of a policy which
has affected the British army since the Victorian period.

At the same time, the differences between the constabulary concept and the guardian force emphasize some of
the particular characteristics of the latter. One of the

fundamental features of the constabulary concept is the
readiness of forces to intervene, but, in comparison, a
basic military weakness of the Victorian army was the
absence of either an élite expeditionary force or a
special reserve to meet the immediate task of speedy intervention. Consistently, the army during the nineteenth
century was caught unprepared, and in times of crisis the
military was rarely able to act without delay. This was
seen even in an expedition such as that to Abyssinia,
where the brilliance of the logistic and support arrangements was marred by the time which it took to assemble the
required forces. For three and a half years the British
Government had negotiated with the Emperor of Abyssinia,
Theodore III, for the release of Her Britannic Majesty's
Consul, Captain Cameron and other European hostages. It
was not until June 1867, however, that the Government,
under pressure from the public, considered military intervention, and after the British Government finally decided
to intervene on 13 August 1867, it was not until 21
October that the advance party landed in Annesley Bay on
the Red Sea to construct the Zulu base, and another three
months before the Commander-in-Chief and his staff
arrived.

The actual campaign was of short duration, lasting
only from January to April 1868, when Theodore was defeated and the captives released. To a large extent,
therefore, preliminary delays were the result of the communications and transport problems which the Victorians
faced rather than any military inefficiency. Before the
extension of the telegraph to the Far East, India and
Australasia in the 1870s, for example, it could take
almost eight months to get a reply from a query sent to
New Zealand, while when reinforcements were needed in
the Second New Zealand War of 1860 it took eighty-two
days to rush out the second Battalion of the 14th Foot
from the Curragh.(19) Even so, the lack of contingency
planning, and the absence, until the Stanhope Memorandum,
of any rational ordering of military priorities, exacerbated the natural difficulties faced by the Victorian
military.

This lack of planning was particularly noticeable in
the use which was made of the Intelligence Department in
the War Office. This had developed from the 'Topographical and Statistical Department' which had been created
during the Crimean War, but although it was reorganized
in 1873 and placed under the Adjutant-General, little use
in the 1870s and 1880s was made of the information which
was collected. There was no idea of using this information in strategic or forward planning. 'The Intelligence

Department was simply a depository of facts for those who cared to have them, but as for drawing up plans of operations or for imperial defence in a wider perspective there was none.'(20)   Its main function was seen to be the preparation, on the outbreak of war, of handbooks for use by officers in the field, a legacy of its original pur- pose as a topographical and statistical department.   Some indications of its defects as a planning department were apparent in the 1880s when it was noted that Count Bronsort von Schellendorf, the German Minister for War, had des- cribed the British Intelligence Department as inferior in every way to the German.(21)   But although Wolseley in 1886 succeeded in prising money from the Treasury for the improvement of intelligence work, the military establish- ment continued to be affected by the lack of planning. The Director of Military Intelligence, Sir Henry Brackenbury, revitalized the old department, collecting a vast amount of information which was published in a series of reports for the use of War Office officials and other departments of government.   Yet the reports, once prepared, were rarely kept up to date,(22) and more importantly, little use was made of them.

The basic difficulty was the inability of the civil authorities to decide upon the purposes for which the army existed.   Any attempt to obtain a decision from the govern- ment on this point met with the rebuff that no benefit could possibly result from such idle speculations and that a policy would be determined by the government 'when the time comes'.(23)   This pragmatic attitude reflected very clearly the latent anti-militarist feelings of some poli- ticians, and it made the preparation of contingency plans an impossibility.

It was also very apparent that a 'defensive' attitude of mind characterized the preparation of intelligence reports.   The Department, it was revealed in the Report of the Royal Commission on War in South Africa, did not prepare schemes of operations.   It simply reviewed and forwarded schemes of defence of particular areas which were sent to the War Office by local commanders who them- selves drew up the plans after weighing up the resources available locally.   Even when the Department did make recommendations for future planning, as in 1896, when it urged serious consideration of the possibility of a Boer invasion of Natal, or in 1898, when it specifically recom- mended that defence schemes should be drawn up for the Cape Colony and Natal, no action was taken by the military establishment.(24)   There were two major reasons for this lack of action.   In the first place, the army lacked the General Staff to whom this responsibility could

have been given. More importantly, it was evident that
there was a complete failure on the part of successive
governments to correlate policy and strategy. This reflected very clearly upon the continued assumption on the
part of the civil power that the army was quite separate
from other decision-making bodies. Both deliberately
and unconsciously the army was excluded from any share in
the making of policy. The military implications of political decisions were completely ignored. It was an attitude of mind which was based on the belief that defence
policy was the servant of foreign policy, not its master,
a thesis which specifically excluded from consideration
the extent to which they were co-partners. The practical
problems to which this could give rise were clearly brought
out by Sir Redvers Buller in September 1899:(25)

> I am not happy as to the way things are going. There
> must be some period at which the military and the
> diplomatic and political forces are brought into line;
> and, in my view, this ought to be before action is
> determined upon — in other words, before the diplomat
> proceeds to an ultimatum the military should be in a
> position to enforce it. This is not the case with
> regards to affairs in South Africa. So far as I am
> aware the War Office has no idea of how matters are
> proceeding and has not been consulted.

The Victorian army thus lacked the readiness of constabulary forces to intervene, for a pragmatic defensive
approach to strategy precluded even the consideration
in advance of offensive operations. All too frequently,
this unpreparedness had tragic results when the army was
eventually ordered to act, but this was part of the price
which had to be paid for the isolation of the military
from other institutions of government.
A comparable contrast between the two concepts of the
constabulary role and the guardian force can also be seen
in their respective attitudes towards 'victory'. In the
constabulary concept, an essential characteristic is that
the military seeks viable international relationships
rather than victory, because it has incorporated a 'protective military posture'. In this context, the attitude
of Victorian officers was ambivalent. Fundamentally, as
an example of a traditional military organization, the
army endorsed more general public demands for victories,
and few officers would have been prepared to suggest that
victory in itself was unimportant. There were exceptions.
Butler in South Africa in the late 1890s sought to recon-
cile the Boers rather than drive them into war, a preference which subsequently led to his being subjected to
considerable and violent criticism both in the press and

in public. But most officers were prepared to drink to
the toast, 'Here's to a bloody war' and to see issues in
uncomplicated shades of black and white. Yet when serving in politico-military appointments, officers, in
identifying themselves with the civil power, tended to
adopt a broader point of view and be prepared to accept
that victory per se was not necessarily the sole objective
of the army. This was particularly noticeable in situations such as the Siege of Chitral in 1895 when, while
trying to resolve a complicated issue of the succession
to the ruler of this minor state on the borders of India
and Afghanistan, British troops were besieged for forty-
six days in the ruler's fort. Here the goal of military
victory was subordinated in importance to the need to
protect the nominated Rajah, a young boy. The maintenance
of viable international relations, in accordance with
British policy in this part of the world, was thus of far
greater importance than the immediate defeat of the
usurper, Sher Afzul.(26)

The Chitral example brings out clearly an important
characteristic of the guardian role, that is, the notion
of protection, primarily a defensive rather than offensive
attitude, and this characteristic is evident in many of
the activities which the implementation of this role
necessitated. Thus in the United Kingdom, the Victorian
army was involved in two main areas of interest. First,
the army had an ill-defined responsibility for the protection of the country against an external aggressor, a
traditional role which was always under attack from those
critics who believed that this task could be more efficiently carried out, at lower cost, by auxiliary forces.
As Dilke and Wilkinson said:(27)

The military function of home defence can be well
fulfilled, not indeed by 'men with muskets', but by
civilians who have acquired skill in arms and a sense
of discipline, and are well-organized, armed and led.

Second, the army was responsible for aiding the civil
power in the maintenance of internal law and order,
another traditional commitment, which although it had
declined in importance after the formation of police
forces, remained to the end of the Victorian period, a
very real task. In Lincolnshire, for example, in 1901,
a lock-out in the fishing industry resulted in the outbreak of riots on the Grimsby docks when the trawler-
owners sought to bring in foreign crews and engineers from
inland towns. Though the fishermen were exhorted by their
leaders to co-operate with the police, detachments from
the Lincolnshire Regiment were moved into Grimsby to maintain law and order.(28) When several thousand free

labourers were brought into Hull in 1893 to break a dock
strike, protection was provided by the military and by
two gunboats which were stationed in the Humber. Similar
military protection was provided in Southampton and
London.(29)
     Outside the United Kingdom, the characteristic activi-
ties of the guardian role were more complex, since they
were modified throughout the nineteenth century to accommodate changes in colonial and foreign policy. Five areas
of activity, however, constituted a permanent core which
persisted irrespective of transient amendments to detail.
First, a major commitment was the traditional task of
guarding and defending 'bases from which the navy could
control trade routes, intercept attack on British territory and in company with the army, act offensively against
enemy territory'.(30) To fulfil this role, troops were
stationed in every part of the world from Bermuda to the
Cape, and from Nova Scotia to Mauritius, and, increasingly,
as bases grew into colonies, this commitment was similarly
enlarged. A second task for the British military was thus
the protection of these enlarged base areas and other
colonial territories from external aggression, a commitment which tied down a large part of the army until, in
the decade from 1860-70, a changed policy which encouraged
the principle of colonial self-help in defence permitted
the withdrawal of many overseas garrisons from the partially or wholly self-governing colonies. By 1870, for
example, Cardwell had carried the policy of withdrawal
from the self-governing colonies almost to completion.
All British troops had left Australia, New Zealand and
Tasmania, and the last were soon due to leave Canada except for the garrison at the naval base of Halifax.
     The reduction, on the ground of financial expediency,
in the number of colonial garrisons meant that in 1870
these numbered less than 24,000 troops (excluding India)
in comparison with nearly 50,000 ten years earlier. This
did not, however, affect a third commitment, which for its
implementation still depended on the use of British
troops in the remaining colonies and protectorates to
maintain law and order. This internal 'policing' role was
again a traditional task in which the Victorian army was
employed in actions as diverse as the suppression of race
riots in British Guiana (1856) or in Barbados (1875-6),
operations against the dacoits of Burma (1886-92), the
Battle of the Eureka Stockade in Australia (1854) and in
larger-scale campaigns such as those in Canada in 1870
(the Red River Expedition) or in 1885 (suppression of
Riel's Rebellion). Irrespective of the moral justification for some of the actions which were taken in

implementing this role, internal policing tasks at least had the merit that the military retained a protective posture which was based on the need to defend rather than to attack. The remaining two commitments which were accepted by the Victorian army could less easily be justified on these grounds, and their implementation suggested that the military was being used offensively as an instrument of expansion. This was particularly noticeable when politicians considered that the goal of protecting traditional British interests could only be achieved by a campaign of conquest or annexation. The extension of the area of 'Pax Britannica' which followed the successful completion of such a campaign, was justified on a number of grounds. As Southgate comments on one such area of expansion:(31)

> In New Zealand, sovereignty was assumed, reluctantly, because there was no choice between settlement and no settlement, only between extra legal uncontrolled settlement and lawful, regulated settlement.

In any one of these situations, the Victorian army was expected as the agent of colonial and foreign policy to respond to the demands which were placed upon it, and as a means of extending British influence, particularly at the time when 'indirect imperialism' was the order of the day, it became involved in a succession of minor campaigns. Equally, the final commitment with which the army was tasked, frequently embroiled it in minor but hard-fought wars. This again was an area in which offensive rather than defensive action characterized the steps which the military undertook, for the castigation of those who had wronged or insulted Great Britain or British subjects demanded the despatch of punitive expeditions. In these, the military equivalent of 'gun-boat diplomacy', the Victorian army was yet again involved in campaigns where the military weakness of the enemy was counterbalanced by difficulties of terrain, logistics and communication.

Frequently these areas of activity became intermingled, so that a punitive expedition, as in Burma in 1885, resulted in the eventual annexation of the conquered territory, but this was not inevitable, and the examples of the Third China War of 1860 or the Abyssinian Expedition of 1867-8, show that annexation was not always the ultimate goal. Similarly, the meaning of 'British interests' could be widely interpreted, so that the 'defence' of a territorial area could be identified with a campaign of expansion. Thus the extension of the Indian frontier to the Indus after the disastrous First Afghan War of 1839-42 was, in part, the result of a British obsession with the need to defend the North-West Frontier of India against possible Russian expansion.(32)

But irrespective of the theoretical distinction between
the attainment of a specific objective or the adoption of
multiple goals, a characteristic common to all these areas
of activity, a characteristic which had a considerable
effect on the development of military professionalism, was
the emergence of a 'small-war mentality'. This is not to
suggest that all the campaigns fought were identical in
form.   Indeed they were of a bewildering variety conducted
against enemies many of whom were far from primitive or
derisory as opponents.   Too many commanders found to their
cost that they could not dismiss out of hand the fighting
capabilities of the 'fuzzy-wuzzies' or 'wogs'.   The Boers,
with the American Indians, were the finest irregular
cavalry in the world.   Like the tribesmen of the North-
West Frontier they were excellent marksmen, masters of the
ambush and tactical withdrawal.   The Sikh army, trained by
Europeans and well-equipped with light artillery was, until
its defeat in 1849, a powerful force of 40,000 men which
always posed a threat to British forces in India.(33) But
these colonial enemies were not the mass army of a technological society;   they were in no way comparable with the
forces which opposed each other at Bull Run or Sedan, and
the campaigns which were fought by a colonial army
differed considerably from wars of attrition and trench
warfare.
These differences between these small wars and conventional warfare were clearly summed up by Captain C.E.
(later Major-General Sir Charles) Callwell in 1896.   He
suggested, on the basis of the experience gained by the
Victorian army in campaigns fought from Canada to the Cape
and from Pekin to Ashanti, that in small wars, 'conditions
are so diversified, the enemy's mode of fighting is often
so peculiar, the theatre of operations presents such irregular features, that irregular warfare must be carried
out on a method totally different from the stereotyped
system'.(34)   Since these were the actions in which, with
the solitary exception of the Crimea, the British Army had
been involved from Waterloo onwards, such differences had
a profound effect on the development of military professionalism.   This was particularly noticeable in the reaction of the officer corps to the study of theory, for it
was considered that this often bore little relationship
to what happened in practice.   Until the publication of
Callwell's definitive study of small wars, there had been
no systematic study of the problems faced by the military
in this area of operations.   One result of this had been
the growth of a purely pragmatic approach by the military
to each set of problems when and as they arose, and, unfortunately, Callwell's book appeared at too late a point

in the development of the Victorian army to amend entrenched attitudes. Since these were primarily derived
from a belief that existing military theory was not pertinent in this, a 'small war mentality' developed in
isolation from the main stream of army thought.

The effect of this on military professionalism was
emphasized by the way in which it encouraged the developments of three competing and often conflicting interpretations of the relationship between theory and practice. In
the first of these interpretations, the classical study of
strategy and tactics continued along traditional lines.
This was the study which was exemplified in Hamley's 'The
Operations of War'. It was a development of traditional
eighteenth-century thought based on the concept of
'cabinet' wars in which conclusions could be evidenced by
reference to historical examples. The theory was valid in
so far as it was concerned with the movement of armies as
thought they were pawns on some gigantic chessboard.
Hamley's chapter headings alone spelt this out very
clearly: 'Chapter IV, Case of both armies forming on a
front parallel to the line of communication with the base',
or Chapter V, 'How the conformation of a base may enable
the army possessing it to force its adversary to form
front to a flank'. His selected case studies enforced
his commitment to the 'art of war'. Thus his example in
Chapter IV was the Campaign of Jena in 1805, his 'important deduction' suggesting again that strategy was an
enlarged war game played with pieces of different colours:
'When two armies are manoeuvring against each other's
flanks or communications, that army whose flank or communications are most immediately threatened will abandon
the initiative and conform to the movement of its adversary.'(35) To illustrate Chapter V he selected Moreau's
campaign of 1800, although Hamley did take into account
more recent examples, such as the campaign of Novara in
1849 between the Sardinians under their King Charles
Albert and the Austrians under Marshal Radetzky, or the
campaign in Virginia in 1861. Novara was thus used to
illustrate the effect of operating on a front parallel to
the line of communication with the base. The 1861 campaign in Virginia was an extended example, initially
evidenced by the campaign of 1796 in Germany, of 'The Case
of Independent Against Combined Lines of Operation'.

This was the theory which was studied at the Staff
College in the later years of the Victorian period by
men such as Henry Wilson, whose carefully annotated copy
of Hamley's work and preserved examples of the set examination questions, emphasize the extent to which this study
was a theoretical pursuit firmly founded on historical

precedent.(36)   There were three important points, however,
which the classical theory did not take into account, an
omission which was seized upon in the second interpretation of the relevance of theory to practice.   First, it
did not take into consideration the extent to which the
control of mass armies was not simply a problem in manoeuvring, but a more basic difficulty of organization and
movement.   One of the lessons which should have been learnt
from the Franco-Prussian War of 1870 was that the Prussian
success began with the movement of 1,183,000 men through
the barracks into the army in eighteen days, but it was
perhaps inevitable that Hamley, as the apostle of the classical school, should have concluded that the predominance
of the German armies 'deprives the campaign of much of its
value as a strategical study'.(37)

Second, classical theory failed to emphasize with sufficient force the one major change introduced by a new industrial era, that is, the concept of war as a struggle to
the death between the entire population of opposing
countries.   Previous wars fought by Britain had rarely
involved more than a small proportion of the available
man-power.   Moore's army in Spain numbered less than
35,000 men, a small force by late-nineteenth-century
European standards.   At Waterloo, only a third of Wellington's command were British troops, for the military zeal of
the British people had been channelled into a variety of
separate forces — the Volunteers, the 'Old Militia' em-
bodied for the duration, the 'New Militia' to be called on
in emergency and the Fencibles — all of which had the
advantage or disadvantage that they were available for home
service only.   British military thought tended to accept,
on the basis of these historical examples and because of
the opposition shown to any type of conscription, that any
campaign fought by the Victorian military never would or
could involve the whole of society.   It was a valid conclusion in the context in which it was made, but it did
not take into account the lessons to be learnt from the
example of the American Civil War, nor did it examine the
philosophy of war expounded by the German Carl von
Clausewitz who argued in his book 'On War' that future
wars would always tend towards extremes.

Third, classical thought failed to accept that the
traditional battlefield, as Correlli Barnett points out,
had disappeared: 'The armies of the past, manoeuvring
tightly under the close command of the general, gave place
to extended fronts — detached groups of men close to the
ground engaged in innumerable small fights.'(38)   Individual mobility, not mechanistic movements learned on the
barrack square, was now required.   Immediate responsibility

for leadership was passing from the general at the head of his troops to the battalion or even company commander. This third interpretation of the relationship between theory and practice was, however, associated only with a few officers who were 'European' in their outlook. These were the men, like Kitchener or Wilson, who were sufficiently foresighted to appreciate that any campaign fought in Europe would be a long-drawn-out war of attrition, that it would be a war of considerable casualties and that it would involve not simply a small army of full-time regulars, but the whole of society.

In the Victorian period, when possible intervention in Europe was not one of the activities to which the military was committed, this 'European' interpretation of strategy and tactics was a rarity. Few members of a colonial army saw any relationship between 'classical' or even 'European' military theory, and what happened in practice. In many colonial campaigns, it was the environment rather than the enemy which posed the problems. Wolseley, for example, saw his campaign in Egypt in 1882 as a problem in logistics rather than in tactics or strategy. In the relief of Kandahar the major difficulties faced by Roberts were not caused by his Afghan opponents, but were the result of the physical conditions faced on their 300-mile march by his force of 10,000 men:(39)

> We are marching day after day through a half-deserted land, with no supports to fall back upon in case of disaster, and uncertain of what lay before us; with nothing but thin tents to shield us from a sun which laughed to scorn 100 degrees in the shade, and with a water supply so uncertain that we never knew in the morning where our camping ground in the evening might be.

Kitchener, in re-conquering the Sudan, took two years to move his force of 15,000 men up to the Nile by stages. It crept forward through grilling sun and dancing glare in an environment which was vividly described by Steevens:(40)

> The true Sudan thirst is insatiable . . . is born of sheer heat and sheer sweat. Till you have felt it, you have not thirsted. Every drop of liquid is wrung out of your body; you could swim in your clothes; but inside, your muscle shrinks to dry sponge, your bones to dry pith. All your strength, your substance, your self is draining out of you. . . .

When newly-arrived officers encountered these unfamiliar conditions, their previous theoretical knowledge was, as Steevens pointed out, of little assistance to them:(41)

> You saw how ill-equipped were the men to put up with it. Their heavy baggage, officers' and men's alike —

had been left at the railhead; over 2,500 men had
come with 700 camels. The tents had arrived, but
they were only just being unloaded from the steamer.
The men were huddled under blankets stretched on four
sticks. . . . There was hardly a shelter in the camp
in which a man could stand upright. . . . Taking the
place as a whole, it was impossible to be comfortable
and especially impossible to be clean. It was nobody's
fault in particular, and in this good weather it did
not particularly matter. It happened not to have
begun stoking up at the time; when it likes it can be
mid-summer in March. When it did begin and especially
if it came to a matter of summer quarters, such a camp
as Debeika was an invitation to disease and death. You
have to learn the Sudan's ways, they say, if you do not
want the Sudan to eat you alive. The British brigade
had to learn.

Similarly, the concept of attrition meant very little
to the practical soldier who was fighting an infantry campaign against adversaries whose resources and organization
were so limited that the ultimate result, despite any
temporary setback, could never be in doubt. Classical
theory made even less sense. Fundamental principles
indeed still had some validity. On the basis that the
main object of all strategical combinations was to bring
superior force to bear at a decisive point and time, the
soldier involved in a small war could endorse the relevance of theory. But it was much more difficult to accept
Hamley's detailed instructions which implemented this
theory:(42)

Strategic movements will be considered as having the
following objects: 1st: To menace or assail the
enemy's communication with his base; 2nd: To destroy
the coherence and connected action of his army, by
breaking the communications which connect the parts;
3rd: To effect superior concentrations on particular
points.

The validity of the third object was self evident, and
to a colonial army which was almost inevitably better
armed than its opponent, this aim was axiomatic. But how
could the first object be achieved in a conflict with Boer
commanders whose base was a lonely farmhouse on the high
veldt? What were the enemy's line of communication on the
North-West Frontier of India, or the jungles of Burma and
Ashanti? And, even if these could be found how could they
be destroyed? The control of jungle paths and forest
trails was as difficult for the Victorian army as it was
for the military in Malaya in 1941 or in Vietnam in 1971.
It was not until the last two years of the Second South

African War that Kitchener, who had succeeded Roberts as
Commander-in-Chief, solved the problem of the Boers' bases.
His brilliant but simple solution was based on the harsh
application of sound strategic principles — the destruction of the base. To prevent the Boer commandos simply
melting back into the farming communities, the farms were
burned, and the families of the commandos moved into concentration camps.

But Kitchener was an atypical officer and to most of
his colleagues the strategic theory taught in the training
schools of the Victorian army repeatedly appeared to have
little relevance in the fighting of small wars. In a
guerilla campaign it was difficult for these men to understand how a detachment commander was supposed to comply
with the ruling which decreed that 'before a commander
enters on a doubtful battle, he must first consider
whether the strategical situation demands one'.(43) Far
too frequently, in a colonial campaign, it was the enemy
who decided when battle should commence. In 1879, for
example, the Zulus lured Lord Chelmsford, the local
Commander-in-Chief, away from his camp at Isandhlwana and
then destroyed it and all who were in it. At Bronk-
horstpruit in 1880, when it was uncertain whether a state
of war existed or not, Lieutenant-Colonel Anstruther with
nine officers and 254 men of the 94th Regiment (the
Connaught Rangers) was marching from Lydenburg to Pretoria
when he was halted by a few Boers. The Colonel was
ordered to turn back and, when he refused, fire was
opened at once by a force of about 1,000 Boers who were
concealed in ambush. The action lasted less than an hour
but when it was over the small British force had been
virtually wiped out, all the officers being killed or
wounded in the first few minutes.

Incidents of this kind are in retrospect only minor
tactical reverses which bear no comparison with the
strategical battles with which the classical or European
theorist was concerned. But in the atmosphere of the
Victorian period, with the undue emphasis which was placed
on the virtue of victory, their importance was magnified
out of all proportion. 'Isandhlwana struck the British
public as if it were a catastrophe like the surrender of
Yorktown; another indication that the late Victorians
were getting a little out of perspective'.(44) Elizabeth
Butler, the wife of Lieutenant-General Sir William Butler,
captured the public's preoccupation with these victories
and defeats of its colonial army in a brilliant series
of paintings. Even their titles caught the imperial mood
to perfection: 'The Remnants of An Army', 'The Defence
of Rorke's Drift', 'Steady the Drums and Fifes' and

'Survivor', which portrayed the solitary return of
Dr Brydon to Jellabad in 1838. Equally, her husband, in
common with writers such as Kipling, Steevens, Henley,
Winter and Henty, contributed very considerably to the
public's belief that 'war' in the Victorian era was the
fighting of nothing more than these minor tactical en-
tanglements. Henley, who died in 1903, was remembered for
his anthology of English poems, 'commemorative of heroic
action or illustrative of heroic sentiment'. G.A. Henty
(1832-1902) had begun a life of imperial adventure as
a hospital orderly in the Crimea, using his subsequent
experiences as a war correspondent, in a number of books
with titles such as 'On the Irrawady: A Story of the First
Burmese War', 'By Sheer Pluck: A Tale of the Ashanti' or
'The Dash for Khartoum'. John Strange Winter, pseudonym
of Mrs Arthur Stannard (1856-1911), was referred to by
Ruskin as 'the author to whom we owe the most finished
and faithful rendering ever yet given of the character of
the British soldier'. Among her hundred or so books
published from 1874 onwards, were titles such as 'Army
Society', 'Cavalry Life', 'A Born Soldier', 'In Quarters',
'He Went for a Soldier', 'Heart and Sword', and 'A Blaze
of Glory'. Imperial poets, too, sought to glorify the
victory and defeat of British troops in these colonial
campaigns. The successful defence of Rorke's Drift, a
small group of houses by a ford, by just over 100 men in
1879 was a particularly favourite topic, commemorated for
example by Ernest Pertwee:

>   Nigh twenty years have passed away
>   Since at Rorke's Drift in iron mood
>   'Gainst Zulu fire and assegai
>   That handful of our soldiers stood;
>   A hundred men that place to guard
>   Their officers, Bromhead and Chard.

The over-emphasis which was placed by both the public
and the Victorian military on the importance of these
battles encouraged many officers in their belief that
there was a definite contrast between the theory of war
and actual situations which, as Callwell subsequently
stressed, were completely different from the traditional
stereotyped concept. The apparent irrelevance of much
that was taught thus did little to encourage the study by
individual officers of that underlying theory which in
other occupational groups was already the pre-requisite
of any claim to professional expertise. This was an
understandable reaction. In most colonial campaigns the
real enemy was not the native troops but the environment,
an area with which the theory of war seemed to be unconcerned. Yet a study of theory would have been invaluable

in providing other than pragmatic solutions to the
problems which were encountered, for, irrespective of
whether the campaign was fought in the jungles of Burma,
the equatorial forests of West Africa or the deserts of
Sudan, factors of distance poor communication facilities
and lengthy lines of communication were common to all.

The advantages of this study were particularly noticeable in two other areas of interest to which Hamley drew
attention:(45)

If that army have a preponderating strength in cavalry,
an open country will suit it best; if infantry be its
chief reliance, a hilly or wooded region, which may
neutralise the enemy's superiority in the other arms;
if artillery, good roads and positions which command
sufficient expanse of the country, will be indidpens-
able to its most effective action. To determine this
point a broad and general survey will suffice. But
a more intimate acquaintance with the topography, and
a knowledge of strategy are required.

Individual variations in the level of achieved professionalism were often brought out in the way in which
officers tackled these two problems of the choice of troops
best suited to fight in a particular environment, and the
need for detailed topographical information about the terrain over which the campaign was to be fought. Indian
officers who had been accustomed to campaigning in a country which was very thoroughly mapped, all of it on the
scale of one quarter of an inch to the mile and most of it
on an inch to the mile, were very aware of the value of
detailed topographical information. The delay in landing
troops on the Red Sea coast to march inland to attack
Emperor Theodore at Magdala was largely caused by Napier's
insistence on landing a reconnaissance party in advance of
the main body to find a satisfactory disembarkation point
and a gap through the Abyssinian mountains. Some British
Army officers were equally careful. Wolseley before
embarking on the Egyptian Campaign of 1882 was obliged to
order an investigation of the proposed line of advance.
Kitchener, who as a young engineer officer had directed the
Palestine Survey, was another officer who although lacking
this experience in India, appreciated the importance in a
successful campaign of accurate topographical information.
Himself a skilful surveyor, he took a large mapping mission
with him into the Sudan, setting them the task of surveying almost a million square miles of hitherto unmapped
territory.

In contrast, too many campaigns were fought in areas
where there was a complete lack of this information. When
Hicks Pasha and his Egyptian troops were annihilated in the

Sudan on their march from the Nile to El Obeid in 1883,
their defeat could be attributed to inadequate or erroneous
topographical knowledge which caused them to lose their
way. Two years later, Sir Herbert Stewart lost his way
on his night march from the Abu Klea wells to the Nile.
Straggling through the desolate waste of rocks whose
surfaces reflected the heat absorbed by day, scores of
camels broke down under their loads and had to be abandoned. Straggling caused such disorganization that when
daylight came the column was more of a rabble than a military formation. In South Africa there was a surprising
neglect of any mapping. At Majuba in 1881, on Colley's
night march to turn the Boer's flank by securing the
mountain top, his rear companies lost their way, while his
lack of information about the top of the summit and the
amount of 'dead ground' on the slopes were important con-
tributory factors in his defeat. The lesson, however, was
not learnt, and twenty years later the British Army humil-
iatingly found that it seldom knew exactly where it was
when chasing the Boers across the veldt.

The exercise of professional expertise in the choice of
troops for a particular campaign equally demonstrated
variations among different commanders. Since few generals
had the opportunity of selecting the terrain which best
suited the troops at their disposal, the 'real' professional was the commander who adapted his forces to the
terrain. Again the Indian Army officer showed how he had
benefited from his experience in a sub-continent where
the environment showed every possible variation of climate
and landscape. Napier, before he took 14,000 men through
unmapped, roadless, arid and mountainous country to attack
Magdala, made certain that he had the means to get his
troops through. So his two batteries of seven pounder
mountain guns were adapted for carriage by mules, while the
battery of twelve pounder breech-loading rifled Armstrong
guns were dismantled into sections and carried on the backs
of forty-four specially imported Indian elephants. In
Egypt and the Sudan, Wolseley and his staff sought to
combine the need for infantry fire-power during their
campaign to relieve General Gordon in 1884-5, with the
mobile advantages of the cavalry. Their improvised solution was the Camel Corps which gave an extra radius of
action to the trained infantryman, cavalryman or marine
who was converted into a camelman. Over the terrain
which had to be traversed, the Camel Corps seemed to be
an ideal solution to the problem, and dressed in their
red serge jerseys, ochre cord breeches and blue puttees,
the volunteer officers led their men from Wadi Halfa to
Dongola, 200 miles up the Nile.(46)

But the lessons which should have been learnt from the resourcefulness of Indian Army officers such as Napier or from the few 'professional' generals in the British Army such as Wolseley and Kitchener, were soon forgotten. In most campaigns the commander was content to use his infantry in the traditional way. Even in Egypt and the Sudan, the bulk of the men marched, in the way in which Kitchener's forces marched 118 miles in the desert between 26 February and 3 March 1898 when en route from Abu Dis to Dibeika. In part this was forced upon commanding generals by a persistent shortage of cavalry, but resourcefulness of the kind shown by Roberts and his Staff in the Second South African War, when they formed units of mounted infantry, was not always appreciated. Although the open plateau of the high veldt was ideal cavalry country, as the mobility of the Boer proved, Roberts was attacked both by traditional theorists and by the exponents of the small war thesis. To the former, his advocacy of mounted infantry ignored the classical premise that in the set battle regular infantry were the key to victory. 'The infantry', wrote Lieutenant-General Sir L.E. Kiggell in the sixth edition of Hamley's treatise, 'never found itself unwelcome to the other arms on the South African battlefields.' (47) Equally, officers all too frequently concluded that innovations such as these were unwarranted attacks on the tactics which had hitherto þeen used with success in their colonial campaigns. Above all, this was an attempt, it was argued, to diminish the traditional deference accorded to the cavalry who had always had an important part to play in the defeat of a native enemy.

While this reluctance on the part of many officers to accept the relevance of strategy and the study of theory to the fighting of small wars reflected more general attitudes in the army as a whole, it was also evident that the small war mentality specifically encouraged the perpetuation of out-moded tactics. These were tactics which were out of date in the context of European wars, but the success with which they were employed in colonial campaigns did little to encourage developments in this field. Initially, the adherence to traditional fighting methods was most noticeable in the continued use of the classical square of the kind which had resisted the charge of Napoleon's cavalry at Waterloo. In European wars, improvements in firearms had completely changed the battlefield scene. The French, having introduced the Minie rifle in 1851 had adopted in 1866 the chassepot, similar in style to the Prussian needle-gun, before changing to the tubular-magazine Lebel rifle in 1885. In Britain, the .45 Martini-Henry introduced in 1871 was followed

by a new .303 rifle with a Lee breech and a Metford barrel
which with its magazine, rapid rate of fire and range of
over a mile considerably improved the fire-power of the
infantry.(48)   Four years earlier in 1884, the German
bolt-action Mauser had been modified from a single-shot
weapon into a repeating rifle by incorporating a fore-tube
magazine, while even the smaller European countries, such
as Turkey, who bought the Winchester repeater of 1873, had
increased the efficiency of their infantry. Nor was the
cavalry allowed to carry on armed only with the sabre and
the lance, for the ten-shot automatic Mauser carbine
eventually introduced into the British Army in 1898 was
an ideal cavalry weapon with its effective range of 600
yards, rapid fire and lightness.

The traditional square was already passing from the
European scene in the face of this increase in the fire-
power of the infantry, and the adoption by armies of
either the Gatling machine gun, first used by British
troops in the Ashanti campaign, or the Montigny mitrail-
leuse used by the French in the Franco-Prussian war of
1870, made it even more suicidal to bunch men together on
the battlefield. In China in 1860, the use by the British
of the newly adopted breech-loading Armstrong rifled
twelve pounder also illustrated the developments which had
taken place in artillery, while the French 'Seventy-Five',
the most advanced field gun in Europe in that only the
barrel recoiled instead of the whole carriage, revolutionized battlefield tactics with its far faster rate of
fire, use of shrapnel and cartridges of smokeless powder.

In colonial campaigns, however, where the enemy was
not equipped with these newly-developed weapons, more
sophisticated versions of the eighteenth-century square
were used with success. Sir Herbert Stewart, for example,
in fighting his way from Korti to the Nile in 1885 used
a trundling flexible square to protect his troops on the
march, and although the square was broken when the Gardner
machine-gun and many of the Martini-Henry rifles jammed,
its use enabled him to reach the water supply at Abu Klea
wells. Similarly, Wolseley in the Ashanti campaign
adapted a mobile square formation even to the dense jungle
of West Africa.

In small wars, use could also be made of volley firing
which had similarly passed from the European scene. When
attacked, the command, 'Halt. Volley fire at five hundred
yards. Fire' could be repeatedly given, the range decreas-
ing as the enemy advanced. It was a text-book manoeuvre,
perfected by disciplined drill on the barrack square,
although in Europe the German Army had by the 1880s
totally abolished both volley-firing and the associated

movement of the mass advance in formal order. But in a
colonial campaign, not only was this outmoded tactic still
used, but the success which it achieved could, as Steevens
showed, be interpreted as justification for its use in
other situations:(48)

> The shooting of the British. It was perfect. Some
> thought the Dervishes were mown down principally by
> artillery and Maxim fire; but if the gun did more
> execution than the rifle, it was probably for the first
> time in the history of war. An examination of the
> dead — cursory and partial, but probably fairly repre-
> sentative — tends to the opinion that most of the
> killing, as usual was done by rifles. From the British
> you heard not one ragged volley; every section fired
> with a single report. The individual firing was lively
> and evenly maintained. The satisfactory conclusion is
> that the British soldier will keep absolutely steady
> in action and knows how to use his weapon: given these
> two conditions, no force existing will ever get within
> half a mile of him on open ground, and hardly any will
> try.

Yet it was not really a satisfactory conclusion,
although it was one which was seized upon by the public
and, more unfortunately, by many members of the Victorian
army. The Dervishes, although the most formidable and
dangerous native enemy, with the possible exception of the
Maoris, which the British ever had to face, lacked the
fire-power of contemporary European armies. Bereft of
artillery, armed principally with the spear, although their
riflemen used a variety of weapons from Remingtons to ele-
phant guns, the Dervishes were not a European enemy. Over-
whelming superiority of, rather than near equality of,
armaments, permitted the retention of the square and volley
firing so that the British and European troops at Omdurman
were drawn up in formal battle array just as though Wellington or Marlborough were in command. The retention of
these obsolete formations in colonial campaigns was thus
indicative of the way in which the innate conservatism of
the Victorian army was further encouraged by the fighting
of these small wars.

The use of these outmoded tactics, although they were
not employed against the Boers who, armed with the Mauser
carbine and Krupps artillery, were a more respected enemy,
also encouraged the retention of a defensive attitude of
mind in which mobility was sacrificed to caution. This
was not an altogether unexpected reaction in view of the
casualties which had been suffered when commanders had
been too reckless. Nor was it certain whether 'mobility'
in the sense in which it was used in the Prussian Army to

denote the man with a rifle working on his own initiative,
was acceptable in an army where the ordinary rank and file
were thought to be incapable of even thinking - 'a mindless
brick in a moving wall of flesh, instantly responsive to
the orders of his superiors'.(49)

So the small war mentality invoked an insistence on
disciplined solidarity where the soldier was drilled to
machine-like movements, but, more importantly, it completely misread the importance of changes in tactics. The lessons to be learnt in particular from the 1870 war of the
effect of improved fire-power on the conduct of battles
were completely ignored. In colonial wars, the innate
superiority of British fire-power and the absence of any
comparable enemy strength made it unnecessary to inquire
too closely what these changes were. Confident of its
ability to defeat any enemy encountered in the five areas
of activity which primarily governed its fighting, the
Victorian army could ignore or dismiss the potential importance of these developments. Pratt in his 'Precis of
Modern Tactics' spelt out the reason for this:(50)

It is quite true that the introduction of improved arms
has produced very considerable modifications in the
method of fighting; but these modifications are not so
much changes as the growth and development of principles
that have been known for hundreds of years. And it is
a most dangerous thing to ignore all experience obtained prior to the introduction of improved arms, for
it is only by a careful study of the development of
tactics that the true direction in which improvement
is possible can be determined.

The tragic feature of this conclusion was that this
reliance on historical precedent discouraged experiment
and inhibited discussion. The value of the machine gun
as an infantry weapon, for example, was largely ignored.
In colonial campaigns it was primarily used to supplement
or replace conventional artillery. Stewart, for example,
in forming his mobile square, placed his Gardner machine
gun in the middle behind the transport animals. In choosing not to use it as a close-support infantry weapon, he
was following a well-established tradition derived from
French practice in the 1870 Franco-Prussian War. The
French, not realizing that they possessed an entirely new
weapon needing a change of tactics for it to be used most
effectively, replaced one battery in each group of three
six-gun artillery batteries, with ten machine guns. Thus
instead of giving to their infantry the advantages of
increased fire-power, the French had, in effect, replaced
part of their artillery with riflemen, for both the machine
gun and the rifle had a comparable effective range of about

1,000 yards, far less than the range of the artillery
which had been replaced. Its potential effectiveness as
a close-support infantry weapon in mass warfare could not
be visualized. Nor was there much discussion about its
future use in the British Army. To Haig, the machine gun
was quite simply 'an over-rated weapon', an attitude of
mind which reflected an orthodox point of view. This
was based on the argument that 'The rifle, effective as
it is, cannot replace the effect produced by the speed of
the horse, the magnetism of the charge and the terror of
cold steel.'(51) The lance was thus re-adopted in 1907
as a cavalry weapon in place of the Mauser carbine, and
this orthodox school of thought, best personified by Haig
and French, continued to stress the importance of the
cavalry role on the grounds that 'bullets had little stop-
ping power against the horse'.(52)

This wish to reassert the dominant position of the
cavalry in a hierarchy of military arms was another facet
of the small war mentality. By its implicit reference to
the relationship between the socially acceptable cavalry
and chivalric concepts of honour, it also drew attention
again to the perpetual importance within the Victorian
army of the trait of 'character'. This was particularly
noticeable in these colonial campaigns, because in the
absence of technological developments which might have
acted as a check, moral attributes assumed an increased
significance. This could be readily rationalized, as
Lieutenant-General Sir Lancelot Kiggell showed:(53)

> History proves to the hilt that in all ages the moral
> has been to the physical as three to one. Courage,
> energy, determination, perseverance, endurance, the
> unselfishness and the discipline that makes combination
> possible — these are the primary causes of all great
> success, and in turning our thoughts to new guns or
> rifles or bayonets, we too often forget the fact.

This was the 'character' rather than 'intellect' which
for too long had been accepted as the fundamental criterion
of professionalism, in areas as diverse as the Sandhurst
and Woolwich training programmes, or the selection of
officers to fill élite appointments on the Staff or in
command. In colonial campaigns, it assumed an even greater
importance than in the army at home, for in these territories, its significance was extended through its association with wider Imperial concepts of 'gentlemanideal'. As
a criterion of military professionalism, its form was
brought out clearly in Steevens' description of Major-
General Archibald Hunter, Kitchener's 'sword-arm' in the
Sudan:(54)

> Reconnoitring almost alone up to the muzzle of the
> enemy's rifles, charging bare-headed and leading on
> his blacks, going without rest to watch over the com-
> forts of the wounded, he is always the same — always
> the same impossible hero of a book of chivalry, he is
> reckoned as a brave man, even among British officers,
> you know what that means.

This was a comment which would have been a generous
evaluation of a young subaltern, tasked with the responsibility of 'looking after his men' and expected to show
evidence of his bravery in battle. But Hunter was no
young subaltern; he was a very senior officer, responsible
for three brigades of Egyptian troops, each of which was
commanded by a perfectly competent brigadier. With subordinates of the calibre of 'Fighting Mac' (Lieutenant-
Colonel Hector Macdonald) or Maxwell who was to put down
the Easter Week rising in Dublin in 1916, Hunter's actions
were inexplicable. His behaviour, however, was in no way
exceptional, since his preference for action rather than
administration was the typical character trait of senior
officers who had been educated and trained according to
the traditions of the Victorian army.

Hunter's attitude also draws attention to another
feature of the colonial army which contributed in no small
measure to the development of the small war mentality and
to the diminution of professional development. These
colonial campaigns laid great emphasis on the personality
of the individual rather than on the system. They were
'subaltern's wars', irrespective of whether command was
in fact delegated to a junior officer in charge of a small
force or whether a senior officer commanded a larger number of troops. In either case a relatively small number
of troops were involved, and the 'character' of individual
officers was a factor of importance in ensuring the homogeneity of the whole. The rational-legal command structure
which characterized the European mass army, where command
was already 'big business', was replaced by an informal
camaraderie in which officers were part of a closed social
circle. From time to time, the newcomer felt very much
excluded from this select group. Sir Ian Hamilton, who
served with the Gordon Highlanders in the 1884 Sudanese
Relief Expedition, was very aware of this when he first
encountered the officers who comprised the Wolseley
Ring:(55)

> Nothing could have been more inhospitable or forbidding
> than the response from the special preserve of the
> Wolseleyites. There was a ban on 'medal hunting'; all
> and sundry were to be properly appointed from London;
> if we wanted to relieve Gordon we must do so via Pall

Mall; anyone attempting to cut in from any benighted
country like India would be summarily dealt with;
arrested certainly; probably tried by court martial...
I was an intruder; I was alone in a hostile camp.
The effects of this attempt to ensure exclusiveness
were subsequently apparent in the larger wars of 1899-
1902 and 1914-18, when considerable strain developed as
an expanded army sought to initiate newcomers to the
officer corps. But this attitude was very much part of
the small war mentality, and perhaps its major effect in
the Victorian period was the way in which through encouraging individualism, it perpetuated a British military
tradition of personal improvisation. Men such as Hunter
or indeed, Kitchener himself, were almost incapable of
delegation. As 'leaders' they were at the head of their
troops; as their own staff officers they tended to keep
everything to themselves, their distaste for and lack of
knowledge of large-scale staff work and organization,
having dangerous overtones for the future.
These future shortcomings, the effects of which were
to be so marked in the mass wars of the twentieth century,
were accentuated in their importance because few officers
in the Victorian military, with the exception of those
who had fought under Wellington in the early part of the
period had any experience of large-scale command. Although a grateful English public was only too prepared to
acknowledge the successes gained by its generals, few
people took into consideration how small were the British
forces which had been involved in these campaigns. The
relative size of the formations which took part in mass
warfare and those employed in colonial wars was often
overlooked. In the Franco-Prussian War of 1870, the
Germans put into motion in the first few days of the campaign, three armies of 409 infantry battalions, each 900
strong, and 354 squadrons of cavalry each of which comprised 150 men. In addition, there were available in
Germany, the 17th Infantry Division and four Landwehr
divisions, giving a total field army, including staffs,
of 462,300 infantry, 56,800 cavalry and 1,584 guns. In
contrast, the largest British force engaged in these small
colonial campaigns was Wolseley's army of less than one
corps and one cavalry division at the Tel-el-Kebir, while
in the whole of the campaign, the troops in the Egyptian
theatre numbered no more than 35,000.(56) Kitchener's
force at Omdurman was even smaller — two British infantry
brigades and four Egyptian and Sudanese brigades, a total
of 13,000 to 14,000 men and 52 guns — but even this total
prompted the public comment that Kitchener, as a junior
Major-General, had been entrusted with a command such as

few of his seniors had ever led in the field. When troops
of the Indian Army were employed outside the sub-continent,
as they were on at least twelve occasions from the Crimea
to Mombassa (1896), no greater numbers were involved. In
the most successful of these campaigns, Napier in Abyssinia
commanded two divisions of infantry comprising four British
and ten native battalions, a squadron of British cavalry
and four regiments of native horse. No commanders thus
had experience of controlling and manipulating large-scale
formations prior to the Second South African War when
eventually some 480,000 Imperial troops were put into the
field.

Victorian senior officers not only lacked practical
experience in this area. They were also short of training
for their positions as military commanders, and the absence
of a formal divisional and corps structure accentuated the
effect of this lack of experience and training on the
development of military professionalism. Army organization
for much of this period was based on the premise that the
military was a loose collection of autonomous regiments
so that it was only for specific overseas campaigns that
higher combat formations were created. The effect of this
had been noted by the Duke of Cambridge before he became
Commander-in-Chief:(57)

No army can be considered as in a proper state to take
the field, however good its component parts may be,
unless it has some organization on a more extended
scale than the mere formations of regiments and batteries, in fact unless a brigade and divisional system be
introduced, which is to be found in every continental
army.

But even when this structure was eventually created,
troops rarely trained together, although in 1872 as part
of a policy which was begun in 1871 and which petered out
a few years later, 23,000 regular troops and 4,000 volunteers and yeomanry did exercise together for a fortnight's
autumn manoeuvres in Hampshire. Higher formations were
primarily administrative conveniences, and as late as the
early twentieth century, the Cavalry Division, scattered
in locations throughout Ireland and England, only trained
twice as a division between 1909 and 1914.(58) This had
two important effects on the growth of professionalism.
It retarded the development of a sense of group homogeneity based on professional rather than social interaction, so that regiments still looked upon themselves as
autonomous units. This was a traditional attitude expressed in a variety of ways, such as the opposition shown
to the disappearance of the old regimental numbers in
1881. 'A very different sense of injury was experienced

when a number was practically lost, and the battle honours of the regiment were merged with those of another perhaps not so variously distinguished corps.'(59)   Its impact was not lessened by this failure to bring together in peace-time those units whose co-operation was a fundamental necessity when war broke out. Second, and more important-ly, this lack of training meant that in the opening stages of any campaign, British troops, who a few months pre-viously had been a collection of independent formations, incurred grave risks of disaster when expected to operate as a homogenous force.

In many ways this failure to create a unified military force was not the result of any deliberate policy on the part either of the civil power or the military élite. To a considerable extent, it was a natural concomitant of a situation in which the Victorian army was always short of troops to carry out even the limited tasks of the guardian role. This was brought out clearly in the 'Edinburgh Review' in 1885:(60)

> With 60,000 men in India, with 16,000 men on the Nile, with 30,000 men in Ireland, with an expedition on its way to South Africa of about 5,000 men and with the Mediterranean garrisons, a mere fraction of the army remains in Britain. . . . It is apparent to everyone who will honestly consider the subject that the present military establishment is not adequate to so great and various a task.

The shortage of troops was a consistent feature of the position in the United Kingdom where the commitment of the army was to national defence and to the preservation of public order. During the years of political, industrial and agricultural unrest in the 1830s, for example, the major source of power available to the Government in their bid to restore law and order, was the Brigade of Guards (4962 officers and men) and the cavalry. Only four infan-try battalions, each of a nominal 800 men, were stationed in the United Kingdom, although since cavalry were more useful than infantry in preserving the peace, the shortage of infantry in this period was no great hardship. At the same time, insufficient regular cavalry were available to meet the task which had been so clearly defined by the Duke of Wellington:(61)

> It is much more desirable to employ cavalry for the pur-post of police than infantry; for this reason, cavalry inspires more terror at the same time it does less mis-chief. A body of twenty or thirty horses will disperse a mob with the utmost facility, whereas 400 or 500 in-fantry will not effect the same object without the use of their firearms, and a great deal of mischief may be done.

Since the Household Brigade, 1,202 strong, and the
cavalry regiments, whose 6,379 men were largely stationed
in Ireland to man 600 garrison posts, were insufficient
to meet this commitment it was the yeomanry who were often
used in these para-military operations. In many respects,
the officer corps did not regret this transfer of their
task to the yeomanry. It had the disadvantage that the
use of volunteer forces encouraged an attitude of mind
which argued that there was nothing done by the standing
army which could not be more cheaply carried out by
auxiliary forces. It also meant, though this was a point
which was not necessarily appreciated by the military
establishment, that the civil authority, by limiting the
number of regular forces kept in the United Kingdom,
coincidentally limited the potential power of the officer
corps as the controllers of an organized force. But para-
military operations were rarely welcomed by the army.
These operations conflicted with their interpretation of
military 'honour', although this did not prevent officers
from carrying out their duties when ordered, for the interests which they were protecting were their own.
In the Highland clearances of the nineteenth century,
for example, one of the proprietors whose actions in
evicting his tenants produced the unrest which led to an
attack by the police on the women of Clan Ross in 1854,
was Major Charles Robertson, 7th Laird of Kindeacre (78th
Highlanders).(62) Even so, officers often found these
duties to be distasteful. Thus, Major-General David
Stewart of Garth, who as an officer of the 42nd Foot in
1792 had been with his troops when they were used to
protect the movement of sheep into Easter Ross, subsequently argued in 'Sketches of the Character, Manners and
Present State of the Highlanders of Scotland' (1822) that
'The manner in which the people gave vent to their grief
and rage when driven from their ancient homes, showed that
they did not merit this treatment, and that an improper
estimate had been formed of their character'.
A very realistic reason for disliking police work,
however, was the knowledge that repeated outbreaks of
unrest imposed a considerable strain on an overtaxed military establishment, by involving more and more troops.
Some of these engagements were in the military sense minor
affairs. At Bossenden Wood in Kent during 1838, for example, the year when the People's Charter was drafted and
issued, eleven villagers, the followers of John Tom alias
Sir William Courtenay, lost their lives in a battle with
the military.(63) Not all these engagements were small-
scale affairs. In the peak year of the Rebecca Movement
in 1843, Colonel Love had to assemble 1,800 men in the

southern counties of Wales to put down the attacks made there on the toll-gates, yet this number was small in comparison with the forces employed against the Chartists. In the main years of Chartist agitation in England, 10,500 regular troops in 1839 and 10,000 in 1842 were stationed throughout the areas of unrest, while in April 1848 7,123 regulars were assembled in London alone.(64) For some regiments 'police' duty in times of unrest was an almost continuous commitment. During the famine riots of 1846-7 in Scotland, companies of the 27th Regiment (Inniskillen Foot) and 76th Regiment were successively used to protect the loading of grain ships in ports such as Burghead, Wick, Thurso and Invergordon, soldiers of the 76th opening fire on the crowd at Wick.(65)

The difficulties of meeting this commitment in the early years of the Victorian period were highlighted by the small number of troops who were available in the United Kingdom. In 1840, for example, at the height of this unrest, of the 103 battalions on strength, 59 were stationed in the colonies, 22 were in India or China where they were involved in the First Chinese War (1839-42) and only 22 battalions, the majority of which were in Ireland, were based at home.(66)   In addition, many of the soldiers who were available were raw recruits, like the youngsters of the 76th Regiment who, under Captain Evans Gordon, marched with fixed bayonets into Wick in February 1847. Inexperienced and untrained in riot duties, they frequently found difficulty in facing the problems which were set them, tending to over react when under pressure, although the civil police were frequently worse, as the 'Massacre of the Rosses' in 1854 proved. But the lack of experience of these troops meant that the need to meet the colonial commitment kept experienced battalions overseas for long periods. Evidence which was presented in 1838 to the Commission on Naval and Military Promotion showed that of the ninety-seven battalions of infantry which had been in continuous existence for the twenty years from January 1818 to December 1837, one battalion had spent one year at home and nineteen overseas. This was an extreme example, but the average length of overseas service was thirteen years and eight months, so that the ratio of overseas to home postings was approximately two to one. In addition, the practice of rotating battalions meant that few soldiers spent more than four consecutive years in the United Kingdom. Similarly, 'new' regiments raised during intermittent years of military expansion were often committed overseas. Of the six new regiments raised in 1824, four by the end of 1837 had served eleven years in the colonies compared with three at home, while the remainder had

served twelve years overseas and two at home. Moreover,
the three battalions on the home strength in 1838 were
all destined for overseas service again in 1840, two to
Ceylon and the third to New South Wales.(67)

   This situation produced a major and persistent problem
for the Victorian army. The use of a relatively large
proportion of available troops to maintain law and order
in the United Kingdom, in conjunction with the number
needed to meet the colonial commitment, meant that any
responsibility of the military for the defence of the
United Kingdom was sadly neglected. Disclosures made with
characteristic bluntness by the Duke of Wellington and
General Sir John Burgoyne that the country was defence-
less with no more than five to ten thousand troops on
hand to counter any external threat, evoked a sharp public
reaction. Palmerston was forced to admit to the Cabinet
that 'this Empire was existing only by sufferance and by
the forebearance of other powers'.(68) The moral domination over European powers which Britain had enjoyed from
1815 onwards had now disappeared. Attempts were made to
revive it. In 'Punch', John Bull was depicted as saying
to his 'American Bullies', in 1863, 'Look here, boys, I
don't care twopence for your noise; but if you throw
stones at my windows, I must thrash you both.'(69) But in
the following year Bismarck ignored Palmerston's hints of
war in protection of Denmark, and General Peel was being
very realistic in admitting to the Duke of Cambridge that
Prussian intervention in entering Holstein could hardly
have been prevented:(70)

   If when the Prussians and Austrians entered Holstein as
   they said only with the intention of preserving peace,
   we could have said; 'Well we highly approve of this
   and we will send 60,000 men and our fleet to the
   Baltic', the gross robbery that was afterwards committed would never have been perpetrated.

This loss of European prestige was the price which had
to be paid for diverting all available troops to man the
colonial army. Repeatedly, in the absence of any rational
ordering of military priorities, the Empire consumed an
ever-growing proportion of the available troops. This
had a number of results. Militarily, it meant that
Wolseley's expedition to Egypt could only be mounted by
calling out the First Class Army Reserve — men who had
passed into the Reserve in 1881 — although the original
intention was that they were to be mustered only in time
of national emergency. It also had the effect that successive Secretaries of State for War sought unsuccessfully
to find a solution to this permanent shortage of troops.
Cardwell tried to solve it by revising an existing system

whereby in a regiment with two battalions, one served
abroad whilst its counterpart remained at home in the
regimental depot as a training and reinforcement unit.
This Cardwell had extended to all regiments but the demand for troops in the colonies meant that by 1882 there
were eighty-one battalions of infantry abroad and only
sixty at home. Moreover many of the latter were below
their full active service establishment of 950 men, and
of the troops available at home, a high proportion were
raw recruits or men who could not be sent overseas on
active service.

The difficulty seemed to be insoluble. The government,
it appeared, had to choose between a strong colonial army
and a strong home army. There was no compromise, for the
problems caused by a shortage of available troops had been
clearly seen during the Crimean War. After the first
25,000 men had been sent to the Crimea, subsequent reinforcements were difficult to find. The first group of
seven battalions consisted of 6,000 men, but a second
group of eleven battalions consisted of 6,500 men, and by
mid-1855 the British Army numbered only about a quarter of
the French total of 90,000 men. In an attempt to attract
recruits into the army, the War Office turned once again
to the Highlands which had been a major source of recruits
during the American Revolution and Napoleonic Wars. But
although the lairds of Sutherland, Argyll, Seaforth and
Gordon were asked to muster their young tenantry in new
battalions of the regiments which carried their names, no
men were forthcoming. In Sutherland, the Duke offered
a bounty of £6 from his own pocket to every man who en-
listed in the 93rd Regiment, the first battalion of which
was under orders to go to Scutari, but not one single
recruit came forward. In the Isle of Skye, which up to
1837 had provided 699 officers and over 10,000 men to the
Highland regiments, including twenty-one general officers,
only one man, subsequently dismissed, was recruited. When
Lord Macdonald sought to persuade his tenants to join the
78th Regiment (The Rosshire Highlanders) he was told to
send his deer, his roes, his rams, dogs, shepherds, and
game-keepers to fight the Russians.(71)

If recruits could not be readily obtained, or if Parliament was unwilling to pay for the necessary expansion
of the military, then it was apparent by the mid-1880s
that the army was in a deplorable condition. This was
brought out very clearly in 1885 when with the army fully
committed in Egypt and the last two available battalions
sent out from Britain, the Russians threatened British
security in India by an advance towards the frontiers of
Afghanistan. It became readily apparent that Britain

could not implement her guardian role both in India and
Africa at the same time. To meet what Gladstone called
'unprovoked aggression' in Afghanistan,(72) it was planned
for troops to be withdrawn from the Sudan and for a line
to be held in Afghanistan from Kabul to Khandahar. But
these were stop-gap measures. In withdrawing from the
Sudan, Lord Rosebery made it clear that the menace in
Central Asia and Afghanistan 'opened an abyss which I do
not like to contemplate'.(73) Britain was reluctant to
yield up Africa to an imperialist France which was assert-
ing her claim to an Empire in North Africa.

What was also apparent was that an economy-minded
government with Childers as Chancellor of the Exchequer
was determined not to increase the military establishment,
preferring instead to take away the last Canadian garrison
at Halifax or even use the Brigade of Guards on foreign
garrison duty, a proposal that was immediately blocked by
the Queen.(74) This was a situation which engendered
considerable civil-military strain, for it was apparent
to the military that the army was being continually
sacrificed to political expediency. Although the military
situation was relieved by the dubiously legal use of the
Indian Army for garrison duty and military expeditions,(75)
this did not alter the conclusions which were drawn by the
military. Indeed, it increased military dissatisfaction,
for this use of Indian sepoys and cavalry made it clear
that the military had not underestimated the seriousness
of the position, even though politicians were not prepared
to admit this in the House of Commons. In addition, the
military was incensed by the attempts of other politicians
who tried unsuccessfully to find a solution to a problem
which was apparently intractable. Most of them began from
the unfortunate premise that the Victorian military was
basically a colonial army, an attitude of mind which was
succinctly summed up by Dilke and Wilkinson. They argued
that the defence of the United Kingdom rested mainly on
the maintenance of a strong Royal Navy, to which military
forces were supplementary. This recognized a thesis
which stressed that, while the maintenance of large
standing armies was appropriate policy for those continental European states which had long land frontiers to
defend, such a policy was inappropriate for the military
requirements of an insular state. Accordingly, the purposes for which the Victorian military establishment
existed, were thought to be, the defence of naval bases,
the defence of India, imperial policing, counter-attacks
on the territory of an enemy after the command of the sea
had been obtained, and the defence of Great Britain
against invasion by a coup-de-main, or after the extreme
case of naval disaster.(76)

On the basis of these arguments, Dilke and Wilkinson
proposed that a 'separate' army could be created for overseas service by extending the existing system whereby a
battalion, though not necessarily its members, could be
stationed overseas for sixteen years. This idea of creating a completely separate colonial army was no novel
solution to the problems created by a persistent shortage
of recruits. It had, for example, been suggested by
Arthur Otway in the House of Commons in 1880, for he had
proposed that the army could be divided into a few 'home'
regiments and a majority of 'overseas' regiments. In his
opinion the need to keep military forces in dispersed
locations throughout the United Kingdom had completely
disappeared. The task of maintaining law and order had,
he argued, been transferred from the military to civil
police forces, and, in those rare instances where the
degree of internal unrest necessitated the use of the
army, the development of an efficient railway system ensured that the required military contingent could be
quickly and easily moved to any trouble spot.(77)
    These arguments were very popular ones. They confirmed
a public perception of the military image. This was derived from the perceived differences between a home-
based army in which ritual and ceremonial appeared to be
the only goals of importance, and an heroic colonial army.
'A man', argued Colonel Henderson 'must have been east of
Malta before he is qualified to sit in judgement on the
regular army of Great Britain. The beardless regiments
of Aldershot or the Curragh can no more compare with the
masses of strong men . . . who hold India and Egypt, than
the lazy routine of English quarters can compare with
the vigilance and stir of the restless East.'(78) They
recognized a feeling that war was an exciting adventure
in which the triumphs of the military overseas could be
repeated in England by a citizen army. Indeed, these
volunteers forces could form, it was argued, the counter-
attacking expeditionary force for 'there would be abundant
material ready to hand in the defensive forces, whose
training, even if they had never been engaged, would have
been improved during the time they were kept under arms.(79)
    It was against this background, in which the regular
army was consistently seen to be a colonial-based establishment, that Edward Stanhope (1840-93) sought to produce
the first ranking in the Victorian period of military
priorities. In his memorandum, the then Secretary of
State for War clearly laid down, in order, the tasks of
the British army:
    (a) The effective support of the civil power in all
         parts of the United Kingdom.

(b) To find the number of men for India, which has
     been fixed by agreement with the Government of
     India.

(c) To find garrisons for all our fortresses and coal-
     ing stations, at home and abroad.

(d) After providing for these requirements, to be able
     to mobilize rapidly for home defence two army
     corps of regular troops and one composed partly
     of regulars and partly of Militia, and to organize
     the Auxiliary Forces not allotted to Army Corps
     or garrisons, for the defensible positions in
     advance, and for the defence of mercantile ports.

(e) Subject to the foregoing considerations and to
     their financial obligations, to aim at being able
     in case of necessity, to send abroad two complete
     Army Corps with Cavalry Division and Line of Com-
     munications.

The primary identification of the Victorian military
establishment as a colonial guardian force, however,
could be seen in the rider which Stanhope added:(80)

But it will be distinctly understood that the probability of the employment of an Army Corps in the field
in any European War is sufficiently improbable to make
it the primary duty of the military authorities to
organize our forces efficiently for the defences of
this country.

In any evaluation of military professionalism during
this period, the Stanhope Memorandum was and is a basic
point of reference. It confirmed the thesis that, in the
eyes of the civil power, the military establishment per se
did not have a monopoly of occupational activity, since
considerable importance was placed on the part to be played
by Volunteer Forces. On the other hand, while Cardwell in
1873 had not accepted the recommendation of the Mobilization Committee that all auxiliary forces should be integrated with the regular army, these forces no longer
operated as completely separate units. The Lord-Lieutenant,
for example, had lost his right to nominate officers to
these forces and they came under the control of the War
Office. Even so it was not until the Victorian period had
ended that a fuller appreciation of the extent to which
the military task was a professional rather than an amateur
occupation led Haldane in 1907 to place these auxiliary
forces more completely under the control of the regular
army. In the Victorian period, therefore, it could not be
argued as it was after 1907 that, 'there was no escaping
the truth that the (Territorial) Force was now an Army
organization, controlled by the War Office, commanded
down to Brigade level and sometimes even lower by regular

officers, and subjected to rules and regulations emanating,
not from the Town Hall or castle, but from an office in
Whitehall'.(81) Indeed a characteristic feature of the
Victorian army was the persistent emphasis which was
placed on the separate importance of auxiliary and reserve
forces.

This, in turn, had an effect on any rationalization of
the goals set for the regular army. Primarily, it was
envisaged that the defence of the United Kingdom would be
one of the lesser tasks of the regular army. 'The Navy',
it was said, 'can deal with home defence.'(82) In this
area of defence planning, the influence of the 'blue
water' pressure group was very pronounced. The extreme
exponent of this school, Admiral Sir John Fisher, went so
far as to argue at a later date that 'The Navy is the 1st,
2nd, 3rd, 4th, 5th . . . ad infinitum Line of Defence. . .
if the Navy is not supreme, no Army, however large, is of
the slightest use.'(83) When to this argument was added
the conviction that the defence of the United Kingdom
could, in any event, be left to the Volunteer forces, it
was evident that the sphere of military action had to be
outside England. Nor in this era of complacency was it
envisaged that this would include intervention in Europe.
Again, the 'blue water' school subsequently made this
quite clear. It deprecated the 'thin edge of the insidious wedge of our taking part in a Continental War, as
apart absolutely from coastal military expeditions in pure
concert with the Navy.'(84) Since there was no suggestion
that the British Army should become involved on the continent of Europe as a result of diplomatic agreements, this
exclusion of the military from a European role emphasized
again the extent to which the army was seen to be a colonial force. As Viscount Esher later argued,(85)

> The Army is maintained for offensive-defensive purposes
> of the Empire. It is not to be organized for the de-
> fence of these shores, but is intended to take the
> field at any threatened point where the interests of
> the Empire are imperilled, and especially on the North
> West frontier of India.

The Stanhope Memorandum thus had the merit that, rightly
or wrongly, it was written in accordance with a traditional
point of view. The priority of tasks which it established
confirmed the imperial orientation by clearly defining the
purposes for which the army existed. If the professionalism of the Victorian army is then examined only against
these laid-down goals, it can be argued that the army was
able to meet the commitments with which it was tasked with
almost complete success. Certainly, there were minor
tactical reverses, but these were outweighed by the

brilliance of other campaigns which brought out very
clearly the army's ability to rival in these colonial
small wars the engineering and technological feats of
Brassey or Brunel in the field of technology. Since it
was the terrain rather than the natives which was the
true enemy, the ability of this colonial army appeared to
be evidenced quite clearly in the construction of enterprises such as the Sudan Military Railway. Steevens
showed how onerous a task this had been:(86)

> Everybody knew that a railway from Halfa across the
> desert to Abu Hamed was an impossibility . . . before
> the end of 1897 the line touched the Nile again at that
> point 234 miles from Halfa and the journey to Berber
> took a day instead of weeks. There was no pause at
> Abu Hamed; work was begun immediately on the 149 mile
> stretch to the Atbara.

This was the achievement of Bimbashi Percy Girouard,
one of three subalterns specially chosen from the Railway
Department of the Royal Engineers who, after his colleagues
died in the Sudan, created a railway in which over 500
miles of track were laid across a savage desert.(87) But
this was not an isolated example of Victorian enterprise.
In the Red River Expedition of 1870, troops moved across
more than 1,400 miles of unmapped Canadian waste to reach
the Winnipeg area. In Abyssinia, an artificial harbour
was built in Annesley Bay before construction started on a
railway from Zula to Koomayli. When the local fresh water
resources proved to be incapable of providing the minimum
ration of one and a half gallons per man each day, con-
densers provided by the Admiralty were set up to remedy
the deficiency. In West Africa, to reach Kumasi, the
capital of the Ashanti, the army constructed a road through
the thick fever-ridden equatorial forest, building no fewer
than 237 bridges, including a 200-foot bridge across the
River Pra.

Nor were the deplorable medical arrangements of the
Crimea repeated. In China during the 1860 War, the
wounded were evacuated to the hospital ships where 'every
comfort that skill and invention could supply was prepared for them'.(88) In the Ashanti campaign, a hundred-
bed field hospital was supplemented by the provision of
four hospital ships and an arrangement whereby the Cape
Town lines transshipped patients to Gibraltar or England.
When Napier went to Abyssinia, the Admiralty provided
three hospital ships and, despite the persistent worry
before the expedition was mounted that a high incidence of
sickness could be expected, the rate of mortality was surprisingly low. For Wolseley's expedition to Egypt in
1882, base hospitals were established in Malta and Cyprus;

in the Sudan, although the wounds inflicted by spears
were severe, the medical services were beyond reproach,
the wounded being carefully evacuated in cacolets on the
baggage-camels to the field hospital at Fort Atbara or
Abeidieh where rows of barracks had been converted into a
hospital.

But these successes showed how far British military
experience and expertise had diverged from European experience and professionalism. Relatively easy victories
gained over numerically superior but inadequately armed
native opponents distorted the validity of assessments of
military professionalism which concluded that this was
satisfactory. In itself this was not unreasonable. If it
could be guaranteed by the civil power that the military
would not be used against any other kind of opponent, then
the use of outmoded tactics by the army or the reluctance
of the officer corps to consider more critically wider
strategical issues, were unimportant factors. But if
defence policy were to be the servant of foreign policy
and not its master, then the likelihood of any such
guarantee having a long-term applicability was remote.
Historical experience suggested that a confused vacillat-
ing foreign policy, as in the Crimea in the 1850s, could
inexorably lead to the outbreak of hostilities irrespective of the ability of the army to fight such a war. The
reluctance of the civil power to abandon its apparent right
to control the balance of power on the Continent, suggested,
as in Holstein in the 1860s, that Britain would never give
up her wish to exercise influence over the fate of Europe.
Expansion of the Empire similarly was bound to lead to
clashes with other powers, should the imperialist policy
of the French, Belgians, Germans, Russians, Japanese or
Italians come into conflict with that of Britain. Fashoda
in 1898 showed how the potential enemy in this context
could be another European power; the First and Second
South African Wars brought out only too clearly the shortcomings of military professionalism when the imperial
opponent was someone other than an ill-armed, badly-
organized native enemy.

Ultimately, the assessment of professionalism within
the Victorian army became inextricably bound up with
wider issues of civil-military relationships. In other
occupational groups, an indicator of their professionalism,
even at this early stage of their development, was their
ability to object to lay evaluation of the work which
they did. Doctors and lawyers, in particular, were able
to reject completely any attempt made by the public to
evaluate their professionalism. But the army was a
bureaucratized profession; a great deal of professional

development had been forced upon it by the state which
was its employer. Successive governments had taken steps
to control the shape and nature of military professionalism, either directly by changing the basis of recruitment
or indirectly by deciding the purposes for which the
occupation existed. Consistently the claim by the officer
corps to the possession of a monopoly, that is of the exclusive right to practise the professional skill, had been
countered by the encouragement given by the State to its
citizen forces. In this situation a military reaction
was inevitable and, when it came, it followed an expected
pattern.

For many members of the officer corps, these dysfunctional consequences of an enforced professionalism in a
bureaucratic organization were of little importance.
These were officers with a very limited sense of career
commitment who looked upon their military service as part
of their socialization process. They were the direct
descendents of the moneyed flâneurs of an earlier period
who, having elected to serve in a fashionable regiment,
carefully avoided through judicious exchanges the incon-
veniences of unpleasant postings. This response pattern
was reminiscent of the traditional military image, which
contrasted so vividly with the heroic image established
by successes in minor colonial campaigns. This was the
image evidence to the public by the emphasis placed on
ceremony and ritual, on the niceties of social life and
on the amateur approach to the study of war, which encouraged the claims of those who supported the Volunteer
movement. Fortunately for the Victorian army, not all
officers chose to adopt these neo-feudal attitudes. For
a second group, their response took another form. They
were prepared to surrender tacitly their authority to
participate in policy-making, provided they were given
autonomy to implement this policy in the field. Their
claim to be regarded as 'professional experts' was
accordingly limited to cover only the area within which
their autonomy was effective. These were the 'fighting'
officers, who were willing to abdicate to the politicians
and civil servants the right to dictate overall strategy,
provided they were granted the right to exercise their
professionalism in tactical planning. Unfortunately there
was no guarantee for officers who followed this pattern of
response that the civil power would always make decisions
on matters of broad policy, a failing which then increased
the inherent strain in civil-military relationships. When,
for example, Kitchener was put in command of the Sudan
Expedition of 1896-9, he was placed under Foreign Office
control and instructed to take his orders from Lord

Cromer, the British Agent-General and virtual ruler of
Egypt.(89) Prior to the battle of Atbara, Kitchener,
whose tactical planning had been exemplary, telegraphed
Cromer for permission to attack. The latter in turn
referred this request to General Grenfell, the Commander-
in-Chief in Egypt, who then referred it to the War Office.
Cromer's eventual advice to Kitchener not to attack immediately but to await his most promising opportunity,
was unimportant. What was much more significant was that
Wolseley, as Commander-in-Chief, subsequently made it
clear to Kitchener that in this type of situation, the
strategical decision also had to be taken by the commander
in the field: 'You must be a better judge than Lord Cromer
or me, or anyone else can be. You have your thumb on the
pulse of the army you command; and you can best know what
it is capable of.'(90)
    This was the field commander's dilemma. If he relied
overmuch on the ability of the civil power to make these
decisions, then he could be castigated as an over-cautious
and timid officer who lacked 'drive' or 'initiative'. Conversely, if he were seen to be usurping the prerogative of
the civil power then he invited adverse comments from the
Government, in the way in which Butler was reproved for
his actions in South Africa, and the way in which Colley's
recommendations to Lord Kimberley about the post-war
settlement of the Transvaal in 1880 brought instructions
forbidding him to offer any terms to the Boers.(91) In
some campaigns a potential solution to this problem was
the despatch of a civil plenipotentiary in company with
the military force. In the Third China War of 1860,
Sidney Herbert, the Secretary of State for War, made the
situation very clear to the force commander, Lieutenant-
General Sir Hope Grant:(92)
    It is left to the discretion of yourself and colleagues,
    naval and military, French and English, all acting —
    which is most important — in concert with the two
    plenipotentiaries (Lord Elgin and Baron Gros) upon whom
    the ultimate question of peace or war must rest. All
    I can undertake is that you shall be honestly and
    heartily supported at home.
    In other small wars, the military commander resolutely
refused to have a political adviser on the expedition.
Sir Robert Napier in 1867 was quite adamant that he would
not accept the presence of an all-powerful Political
Officer on the expedition to Abyssinia. His reaction, in
addition, shows a third type of response to these problems
of civil-military relations, for he sought to expand his
effective power beyond the area of traditionally recognized military expertise. 'I respectfully submit', he

wrote to the Governor of Bombay, 'that the political
responsibility should be included with the military command.'(93)   The power of the military in this area was
similar to that of other professionals, in that they continued to exercise a limited veto over proposals that could
only be put into effect with their co-operation.(94) This,
however, was not at this time a dangerous threat to the
civil power, for, essentially, it was a negative approach
which depended for its efficacy on the personality of the
objector.   In the Victorian army there were very few
members of the officer corps whose claim to expertise was
so readily acknowledged outside the army that their presence in a campaign was considered to be indispensable.

A situation which was potentially much more dangerous
was derived from the fourth and final type of response
to these dysfunctional consequences of military professionalism.   This arose when members of the officer corps
sought to redefine the concept of professionalism so as
to include political and administrative skills.   This
went beyond the single wish to exert control over strategic planning, for its implementation would have involved
the officer corps in all areas of decision-making as part
of an 'In-group' who were at the centre of power politics.
In many ways, the attitude of these officers confirmed
the hypothesis subsequently put forward by Finer that it
is the professionalism of the military which can bring it
into collision with the civil authorities.(95)   These
were officers who found the combination of civilian interference, the parsimony of governments, and political
attacks on the military establishment, an intolerable
combination.   Their reaction was most marked. Wolseley, in
a censure motion in the House of Lords on 4 March 1901
after he had retired as Commander-in-Chief, argued that
the existing relationship between the head of the army and
the civil power had to be changed 'to preserve the efficiency of the army and the highest interest of the
Empire'.   In 1905 the Viceroy of India, Lord Curzon, was
so incensed by the attitude of Kitchener, his Commander-
in-Chief, that in a letter to the Secretary of State for
War he argued that Kitchener's proposal for army reform
'would have the effect of establishing a "military des-
potism" which would "dethrone the Government of India
from their constitutional control of the Indian Army"'.(96)

It was perhaps significant that this type of response
was inevitably associated with officers whose military
reputation had been gained primarily, if not exclusively,
outside Europe in the colonial army.   The exploits of
these Victorian heroes had been warmly received by the
public.   Until the Boer War exposed the inherent weaknesses

of the British military system, it had been able to implement, almost with complete success, its responsibilities
as a guardian force. A false evaluation of military professionalism, based on these victories in small wars
against a variety of native opponents, created a dangerous
impression that modern warfare was a distant adventure in
which casualties were minute and success inevitable. It
was only when failure destroyed this sense of complacency
that the military reacted. Then, in the same way that the
public interpreted failure as a symptom of professional
ineptitude, the officer corps sought to put the blame
for their short-comings onto the political power. Ultimately, therefore, the evaluation of military professionalism brought into question not only the ability of the army
but also more general issues of civil-military relationships, and in this area of controversy, one of the most
important issues was the political role of the officer
corps in the parent society. The relationship between the
Victorian army and society therefore cannot be considered
solely in terms of the army's professionalism. Clearly
the latter was of considerable importance, since an
evaluation of the military on the basis of its success or
failure considerably affected the attitude toward the army
of the public at large. But enforced professionalism
did not develop in a vacuum, and military apologists who
sought to blame only the politician for failures such as
the Crimea or the Second South African War, conveniently
overlooked the importance of the army's political role.
This is not to suggest that the Victorian military occupied a central position in the political system similar to
that subsequently enjoyed by the officer corps in Germany
or Japan. Yet the nineteenth-century British officer corps
was not entirely divorced from political participation, and
the relationship of the Victorian army to the parent society was very much affected by the way in which access to
the power centre by officers produced specific military
attitudes towards the British political élite.

7

The Army and its Political Attitudes

The persistent comment that the Victorian army was led
by an amateur officer corps was not only a criticism of
its status as a profession but was also a pointer to the
causes of its political attitudes. This was an army
which, despite the criticisms that could be made from time
to time of its professionalism, still possessed a marked
superiority of arms and organization over any other institution. Consistently, the military was in the forefront
of technological development, yet, despite the material
advantages which it enjoyed, there was little evidence to
suggest that the army posed a threat to the civil authority. The officer corps was content to accept for most of
the nineteenth century a subordinate position, and in
many ways the example of the Victorian army appears to
bear out the contention that it is primarily a fully
professional military committed to the maintenance of its
monopoly over a given area of action, which is a positive
threat to the civil power. The military had the physical
ability to intervene in civil affairs, but it lacked the
motive for intervention, and although there were indications toward the end of the period that increasing military hostility toward the civil power weakened the validity
of this conclusion, these indications are, by their rarity,
a comment on a relatively persistent situation.

The attitude of the military toward the civil power and
the consequent absence of any motive to intervene in civil
affairs, was the result of a complex pattern of both individual and group behaviour and external events. The frequently reached conclusion that the officer was a flâneur,
a dilettante with little sense of career commitment, suggested that he was apolitical because he was not preoccu-
pied with the idea of military 'interest'. Other officers,
who primarily identified themselves with the landed interest of which they were a part rather than with the

military establishment, were equally unconcerned with
political issues which affected the army alone. As a
group, the army did not engage in aggregation or articu-
lation of interests in a manner normally associated with
political parties or pressure groups. Throughout the
period, the army was, in reality, a collection of indi-
vidually distinct regiments, and a large number of officers identified themselves not with the army as such, but
with the cavalry, or the hussars or the artillery or the
engineers or the umpteenth foot. Indeed, the distinction
between the Scientific Corps and the Army in the early
part of the period, and the subsequent separation of the
home army and the colonial army accentuated these divisions in the military as a whole. But another reason
which was of the greatest importance in shaping the attitudes of officers was the extent to which individuals
initially found themselves able to participate in
decision-making activities. These gave them a ready
access to involvement in politics, in the sense that
politics was concerned with 'who gets what, when and
how',(1) and this easy ingress to various power-centres
greatly affected their attitudes.

This involvement was particularly noticeable outside
the United Kingdom. The requirements in an ever-expand-
ing Empire for an increasing number of skilled administrators who could coincidentally control political development,
created a large number of opportunities for individual
officers. They assumed this power in countless numbers and
at all levels of responsibility. Army Lists of the period
were filled with the names of officers serving in a wide
variety of civil appointments. There were officers training the local militia in Honduras. Kitchener with another
108 officers formed part of the Egyptian Army. The
Director of the Egyptian Slavery Department after 1898 was
Captain Arthur McMurdo (1861-1914), an officer of the 71st
Highland Light Infantry, who was the younger son of
General Sir M. McMurdo. Another officer was physician to
the Crown Prince of Siam, while Major-General Sir John
Ardagh (1840-1907) was the Commissioner for the delimita-
tion of the Turco-Greek frontier.

Senior officers frequently took over the administration
of colonies. Wolseley in Cyprus, Butler in South Africa,
Denison in New South Wales, and Lanyon in the Transvaal
were examples of many officers who filled, with varying
degrees of success, these senior civil appointments. At
lower levels of responsibility, officers filled a variety
of posts ranging from pro-consular appointments to minor
Civil Service posts. In India, Captain John M'Nair, RA
(1829-1910), who had followed a family tradition by

joining the army, was from 1857-68 Comptroller of Indian
Convicts. Subsequently as the Colonial Engineer in
Singapore, he built Government House, gaols and hospitals,
before he crowned a distinguished imperial career by offi-
ciating as the Lieutenant-Governor of Penang for four
years. In Australia, Colonel Henry Pilkington (1857-1914)
of the 21st Hussars was private secretary to the Governor
of Western Australia. In the Sudan, Major Lloyd of the
2nd Scottish Rifles was the provincial governor of a
territory larger in area than England and Wales from 1893
until he rejoined his regiment in 1913.

In appointments such as these, the ambitions of individual officers could be readily satisfied. They changed
with comparative ease from military to civil posts and
back again, their willingness to accept a wide variety
of duties often reflecting a firmly held belief in the
concept of 'Service to the Empire'. But more importantly,
the opportunity to accept these appointments ensured that
officers in no way felt isolated from the civil power. As
governors, consuls, comptrollers and administrators, they
were an integral part of the parent society, and a complex
social network often strengthened by family ties, linked
these members of the Victorian army with their civilian
counterparts. It was this interpenetration of the civil
and military power in these overseas appointments which
gave to many officers their first experience in the exercise of civil authority, and it was a significant factor
in shaping their political attitudes.

For other members of the officer corps, however, a much
more important means of access to the 'centre' of decision-
making was through their participation in national politics.
This occurred at the two distinct but closely related levels
of membership of the House of Lords, and of the House of
Commons. In the first case, a consistent percentage of the
House of Lords during the nineteenth century were military
peers. When Victoria came to the throne, 58 of the 548
members of the House of Lords were coincidentally army
officers. In 1853 this proportion had increased to fifteen
per cent with sixty-three peers in a numerically smaller
House. Towards the end of the century the figures increased still further, until in 1885 and 1898 132 and
182 officers formed about 27 and 35 per cent, respectively,
of the total membership.(2) The contribution of individuals both to the military and to the House of Lords,
however, showed considerable variation, and the statistical relationship between these two groups does not bring
out fully the nature of this contribution. Some peers,
among whom were two of the most infamous officers of the
Crimean War, the 7th Earl of Cardigan and the 3rd Earl of

Lucan, had spent a lifetime of service in the army.
Others were less senior officers who like the young
Philip Sidney Foulis, 2nd Baron De L'isle and Dudley,
had just embarked in 1853 on a military career in a
fashionable regiment. For a number, their service in the
army, unlike that of the peers such as Hardinge, Raglan,
Napier, Wolseley and Roberts who had been ennobled for
their service to the Crown, was a short intermission in a
lengthy socialization process. These were the peers who
spent two or three years in a fashionable regiment in the
way in which others of their contemporaries tasted life
at Oxford or Cambridge, before they settled down to manage
their country estates and to assume the national and local
responsibilities expected from them.

Similarly their contribution to proceedings in the
House of Lords were far from uniform. Only one military
peer had become Prime Minister, although the situation in
which Wellington assumed this office whilst coincidentally
serving as Commander-in-Chief, was a constitutional
anomaly which would have been unheard of in an earlier
era. Others had filled a variety of ministerial posts in
both Tory and Whig governments, in the way in which
General Viscount Beresford had been Master-General of the
Ordnance, Lieutenant-Colonel the Marquis Conyngham had
been Melbourne's Postmaster-General in 1834 and again in
1835, and the 5th Duke of Richmond (1791-1860), a General
and a Waterloo veteran, had not only been a Cabinet
Minister in Grey's administration of 1830-4 but had also
taken charge of the Reform Bill in the House of Lords.
Later in the century, this tradition of participation at
the very centre of decision-making was continued by military peers such as the 16th Earl of Derby (1841-1908) or
Lord Wantage VC (1832-1901). Other officers were truly
'backwoodsmen' peers, forming part of the one-third
of the peerage in the nineteenth century who took no part
in political activity. But irrespective of the precise
nature of the contribution made by these military peers to
the government of the country, they formed a part of the
institution of the House of Lords, contributing an essential element to the dignity of the British constitution.
Their political activity supplemented their involvement
in society as a major part of the landed interest. As
Guttsman points out:(3)

> As a group, the nobility and the gentry is further
> characterized by a predominant share in the exercise
> of political power, accruing to them in their totality
> via the House of Lords, and individually through their
> territorial influence and the patronage which they
> exercised and obtained.

The willing subordination of this part of the officer
corps was thus encouraged by their ready access to the
corridors of power, but it must be acknowledged that these
peers formed only a small part of the military at any one
time. Their influence in the army however, because of the
complex family network which ensured that a large part of
the officer corps in general, and the military élite in
particular, were 'aristocratic' and thus related to them,
was very considerable. These ties of 'kinship' were particularly important in a period during which the concept
of patronage and its subsequent legacy, ensured that officers would enjoy the privileges ascribed to members of the
ruling class. But in terms of direct political participation, the avenue of entry for other officers was through
membership of the House of Commons.

This link between the military and the Commons was not
only a characteristic of the Victorian period. In the
Elizabethan House, some half a dozen members had made a
career in the naval or military service, and from the
eighteenth century onwards the military formed the largest
single occupational group in the unreformed House. In the
hundred years from 1734 to 1832, about one in six of the
total membership of the Commons was drawn from the army.
(4) From a total number of 5,134 individuals who represented constituencies during this period, it has been calculated that 847 had at some time held an army commission
and that of these some two-thirds were career officers.(5)
Their numbers in successive Houses remained fairly constant. In 1734 and 1741, when the size of the military
interest in the Commons was at its smallest, they totalled
at least forty-nine. Gradually during the remainder of the
eighteenth century their numbers rose until in the last
Parliament of this century from 1796-1800, the officer
corps provided sixty-five members to a total of 558. The
size of this contribution did not seem to be affected by
the fluctuations of war, for in 1808 as is shown in Table
18, the number remained at a fairly high level. Nor did
the Reform Act of 1832 produce any radical changes in a
pattern which was by this time fairly well established.
As this Table demonstrates, the number of officers in the
first Parliament to be elected after the Act totalled
sixty-seven, a figure which was, in fact, an increase on
the number of forty-seven who had been returned to the
House of Commons in the election of 1831.

Twenty years later, in the period shortly before the
outbreak of the Crimean War, seventy-one of the members of
the Commons had been regular officers, many of whom had
combined a political and military career thus continuing
the eighteenth-century tradition which recognized the

TABLE 18   The military interest in the House of Commons
in 1808 and 1833
Parliament of 1808

Military rank	Kin of peer	Baronet	Son of Baronet	Gentry	Total
General	2	-	-	2	4
Lieutenant-General	7	1	-	3	11
Major-General	5	-	2	5	12
Colonel/Lieutenant-Colonel	20	-	2	11	33
Major	3	-	-	-	3
Junior officers	12	-	-	4	16
Total	49	1	4	25	79

Parliament of 1833

Military rank	Kin of peer	Baronet	Son of Baronet	Gentry	Total
General	1	-	-	2	3
Lieutenant-General	2	1	-	3	6
Major-General	5	-	-	2	7
Colonel/Lieutenant-Colonel	11*	-	1	6	18
Major	3	-	-	1	4
Junior officers	13	-	-	16	29
Total	35	1	1	30	67

Sources: For 1808, Joshua Wilson, 'A Biographical Index
to the Present House of Commons', London, 1808. For 1833,
'Burke's Peerage and Baronetage 1833', and 'Dod's Electoral
Facts'.
* including one Irish peer.

acceptability of this dual role. In addition, the House
of Commons of 1853 showed very clearly the inter-relation-
ship of the Lords and Commons, for many of the officers
who were in the Commons were the kin of those in the Upper
House. This reflected the more general control of the
House of Commons by the Lords. This may have been

diminished in 1832 but through relatives, younger sons
and influence, the peerage continued to maintain after the
Reform Act a direct and active interest in the members of
the Lower House. Writing in 1867, Bernard Cracroft argued
that 326 members of the House of Commons were 'aristocratic', that is they were connected with the peerage and
baronetage by membership, marriage or descent. These
members, he concluded, formed a vast cousinhood so that
the 'Parliamentary frame is kneaded together almost out of
one class, it has the strength of a giant and the com-
pactness of a dwarf'.(6) This was very noticeable in the
officer corps. The Hon. Robert Edward Boyle (1809-54),
for example, second surviving son of the 8th Earl of Cork
and Orrery, was first elected without opposition for Frome
in 1847 as a Liberal. At this time, Boyle, a Lieutenant-
Colonel in the Coldstream Guards, was Groom-in-Waiting to
Queen Victoria, a post he was to hold until March 1852.
As a member of the landed interest, his military and
political roles were complemented by other élitist
appointments as Secretary to the Order of St Patrick, and
as State Steward to the Lord Lieutenant of Ireland. The
8th Earl who was a General and who sat in the House of
Lords under his English title of Baron Boyle, exercised
considerable territorial influence from his seat at
Marston Biggott, Frome. This he used to ensure the return
to the Commons in 1856 as MP for Frome of his grandson,
Lieutenant-Colonel Hon. William Boyle (1820-1908), who as
a Major in the Coldstream Guards had served in the Crimea
at Alma, Inkerman and Sebastopol.
    Irish landowners on a large scale, the Boyle family
were consummate Liberal politicians, the 9th Earl (1829-
1904) who was a Deputy Speaker of the House of Lords,
being appointed Master of the Buckhounds on three occasions
and Master of the Horse twice during the years from 1866
onwards. A Tory family which showed a similar link between a military peer and military members of the Commons
was that of the 5th Duke of Richmond. Here, Lord
Alexander Gordon-Lennox, third son of the Duke and grandson of Field Marshal the Marquis of Anglesey, was returned
to the Commons without opposition in December 1849 as the
Conservative member for Shoreham. Lord Alexander at this
time was a Captain in the Royal Horse Guards in which he
had first been commissioned in 1842 as a seventeen-year-
old cornet, and in the Commons he joined his older brother,
Lord Henry Gordon-Lennox, who had been elected to represent Chichester in 1846. Lord Henry was a Lord of the
Treasury from March to December 1852 and in accepting
government office he followed the example set by their
father, an ultra-Tory General, who had been a member of

Grey's cabinet in the 1830s. The heir to the title, the
Earl of March (1818-1903) who succeeded as 6th Duke in
1860, also enjoyed, as did his brothers, a dual role as an
officer and a politician. A former Guards officer, the
Earl in the 1850s was aide-de-camp both to the Duke of
Wellington and to Viscount Hardinge, his successor as
Commander-in-Chief, whilst coincidentally representing the
county constituency of West Sussex for which he had first
been elected in 1841. Subsequently when 6th Duke of Richmond he became a cabinet minister in Disraeli's administration, although the political act for which he was perhaps
best known was his opposition in 1872 to the Secret Ballot
Bill introduced by Gladstone.
     Not all of the seventy-one military members of the
House of Commons in 1853 were serving on the active list,
and it can be argued that this weakened the potential
power of the army interest as a political force. Many
of the officers had in fact retired or had sold out to
take up a political life as their second career. Lord
Charles Pelham-Clinton was typical of many more officers
in that he had retired in 1851 after eighteen years' service in the Life Guards, during which period he had been
the unsuccessful candidate for Sandwich in 1847. In May
1852, however, he was returned for this Borough, which was
generally accepted to be under the influence of the
Government, as a Conservative who was 'opposed to free
trade but who would not re-impose the Corn Laws'. In
Parliament, he joined two of his brothers. The younger,
Lord Robert Pelham-Clinton, was the member for North
Nottinghamshire, while in the Lords the family was repre-
sented by its head, the 5th Duke of Newcastle who was
Aberdeen's Secretary of State for the Colonies and War.
The military and political tradition was carried on by
Lord Edward Pelham-Clinton (1836-1907). Second son of the
5th Duke he was an Ensign in the Rifle Brigade at
Sebastopol when his father was Secretary of State for War.
A captain in 1857, Lord Edward combined a military career
in which he saw service in Canada and India with a parliamentary career as MP for North Nottinghamshire (1865-8).
On retiring from the army in 1880, he was appointed Groom-
in-Waiting to Queen Victoria (1881-94) and was Master of
the Queen's Household from 1894-1901.
     Similarly, officers retired from the army to seek
election for Irish and Scottish seats. Thomas Bateson,
the member for Londonderry County, had retired on half pay
from the 13th Light Dragoons to succeed his deceased
brother in the House of Commons. The heir of Sir Robert
Bateson who had represented the constituency from 1830 to
1842, Thomas was first returned without opposition for the

county in 1844, and he held on to the seat in the 1852
election. One of the Scottish MPs was George Dundas of
Dundas Castle, Linlithgow. As a young officer, he served
with the Rifle Brigade in Bermuda, Nova Scotia and the
Mediterranean, but he sold out from the army when he was
twenty-five, and three years later in 1847 he was elected
without opposition as a member for the county. Even before
this time, however, the young Dundas who was the heir of
James Dundas, chief of the clan, had been confirmed in his
membership of the landed interest for he had been made a
Deputy-Lieutenant of the county.

Of these officers, twenty-five were still on the active
list while another nine were on the retired or half-pay
lists. Eight of the former were generals, and while all
the remainder were relatively senior field officers, the
most junior officer, the twenty-five year old Hon. C.S.B.
Hanbury who was a lieutenant in the Life Guards, had not
been commissioned until 1850. The generals had often
served in the Commons for some time, and the doyen,
General Sir George de Lacy Evans (1787-1860), had first
been elected for Rye in 1831. An obstreperous Radical
from an Irish landowning family, Sir George subsequently
defeated the Whig candidate Sir John Cam Hobhouse at a
bitterly contested election for Westminster in 1833, and he
continued to represent his constituency from 1833-41 and
from 1846 onwards. At the same time, this Waterloo
veteran pursued a military career, and in 1835 he was appointed to command the British Auxiliary Force in Spain,
an appointment which also brought him a General's commission in the Spanish Army, although he was still effectively the Member of Parliament for Westminster.

Table 19 presents in more detail the background of those
officers who were still on the active list, and from this
and from a study of the wider range of military membership
of the House of Commons, certain tentative conclusions can
be drawn which are more generally indicative of the political activities of the officers. While in the eighteenth
century it was argued that officers sought membership of
the House of Commons because it was the known way to military preferment,(7) no such reason can be put forward
to explain the political motives of the Victorian army.
Involvement in politics in the Victorian period was more
likely to be linked to civil advancement, a conclusion
which was pungently summed up by the General the Marquis
of Londonderry in 1837:(8)

The Statesmen of the present day seem not to know that
a body acting together must have the rewards of ambition, patronage and place always before their eyes and
within their expectation and belief of grasping, as well

as the fine expressions of love for their country, and
the patriotism which is a virtue.

For some officers, this clearly expressed their feelings about a political career. As members of the military,
they served the Crown for a variety of reasons. Family
tradition, a search for status, a genuine love of their
country, patriotism and a belief in the concept of service
were some of the factors which led them to purchase commissions in the army. But a military career rarely led
directly to material advantages. Certainly there was
little financial reward in army service, or any great
amount of tangible rewards in a period of military re-
trenchment. Civil careers promised much greater individual preferment, a factor which led many officers to accept appointments within the Empire. For many officers,
a very realistic appreciation of these advantages made
membership of the House of Commons an attractive prospect.
Some officers assiduously sought nomination to the Commons.
At Windsor, where John Ramsbottom, a former officer of the
16th Dragoons and a local brewer and banker, had shared
representation in the Borough with the Court interest from
1810 onwards, the impending vacancy which would arise on
his death in the mid-1840s prompted Bonham, the Tory
party's agent, to write to Peel, 'My friend Colonel Reid
of the Life Guards and therefore connected with the Court
as acting in his turn as Silver Stick is anxious to fulfil
an intention announced more than two years ago of offering
himself as a Candidate'.(9) It is, however, impossible
in retrospect to determine the precise motivation of these
officers who eagerly put forward their names as Parliamentary candidates. Indeed, the only conclusion which can
be drawn is that it was the deviant, such as General T.
Perronet Thompson, who deliberately set out in a political
career to challenge the civil power. For the majority of
officers, their wish to enter the House of Commons was the
result of considerations of individual advancement which
were closely intermixed with the concept of family
advancement.

The second reason which brought officers into the House
of Commons was thus family interest. This was probably in
the early part of the Victorian period the most important
single reason for the perpetuation of the link between
officers and Parliament. Although the pocket or proprietary boroughs had been affected very considerably by the
provisions of the Reform Act in 1832, the new electoral
conditions in the small boroughs often did little to
diminish traditional family links. Many owners of these
boroughs continued to regard them as being primarily a
means of entry to Parliament for those members of the

TABLE 19   Military membership of the House of Commons 1853:   serving officers only

Name and Family	Commissioned	Regt	Army rank	Seniority	Constituency	Politics	Notes
Hon. George Anson (b 1st Earl of Lichfield)	1814		Major-General	1851	S. Staffs	L	C-in-C East India 1856
Hon. Hugh Arbuthnott (s 7th Viscount Arbuthnott)	1796		General	1853	Kincardine-shire	C	Col. 38 Foot: DL for Kincardine
Sir George Berkley (es Admiral Hon.Sir George Berkley)	1802	RHG	General	1852	Devonport	C	Surveyor-General of Ordnance 1852. Brother of MP for Cheltenham; cousins, MPs for Bristol, Cheltenham and Gloucester
Sir H.R.F. Davie (s Robert Ferguson, MP for Raith)	1818		Colonel	1841	Haddington	L	Major-General 1854
Sir De Lacy Evans (s John Evans,Esq.)	1807		Major-General	1841	Westminster	L	Col. 2 Foot; Lieutenant-General 1854
Sir John Fitzgerald (s Edward Fitzgerald MP)	1800		Lieutenant-General	1841	Clare	L	Col. 18 Foot

Name							
Rt Hon. George C. Forester (2s 1st Lord Forester)	1824	RHG	Lieutenant-General	1848	Wenlock	C	Groom of the Bedchamber; Comptroller of the Royal Household, 1852. Brother-in-law of Hon. George Anson
John Hall	1817	LG	Colonel	1846	Buckingham	C	Major General 1854
Hon. C.S. Hanbury (2s Lord Bateman)	1850	LG	Cornet	1850	Hereford-shire	L	Lieutenant 1854. Fellow of All Souls, Oxford
Lord Hotham (gn Admiral Hotham)	1810	CG	Major-General	1851	Yorks East Riding	C	MP for Leominster 1820-41
R.N.F. Kingscote (gs 6th Duke of Beaufort)	1846	SFG	Captain	1846	Gloucester West	L	ADC to Lord Raglan in Crimea
Hon. W.S. Knox (2s 2nd Earl of Ranfurly)	1844	85F	Lieutenant	1846	Dungannon	C	Groom in Waiting to the Queen; Major 1855
Robert Laffan (n Archbishop of Cashel)	1837	RE	Captain	1846	St Ives	C	Inspector of Railways 1847-52
Hon. James Lindsay (2s 7th Earl of Crawford and Balcarres)	1832	GG	Lieutenant Colonel	1846	Wigan	C	DL for Lancashire: Col. 1854
Henry Lowther (es Colonel Hon. H.C. Lowther MP)	1841	LG	Captain	1843	West Cumberland	C	3rd Earl of Lonsdale, 1872

Name and Family	Commissioned	Regt	Army rank	Seniority	Constituency	Politics	Notes
Lord George John Manners (ys 5th Duke of Rutland)	1840	RHG	Captain	1847	Cambridgeshire	C	Brother of MP for Colchester
Hon. James Maxwell (3s 6th Lord Farnham)	1834	59F	Major	1846	Cavan	C	Severely wounded in Crimea; Lieutenant-Colonel 1854
Lord George Paget (5s 1st Marquess of Anglesey)	1834	4 Lt. Dr.	Lieutenant Colonel	1846	Beaumaris	L	Brother of Lord Alfred Paget, MP for Lichfield; Cavalry Commander in Crimea
E.W. Pakenham (es Lieutenant-General Hon. Sir Hercules Pakenham)	1837	GG	Captain	1837	Antrim	C	Magistrate and DL for Antrim
Jonathan Peel (s Sir Robert Peel)	1815		Colonel	1851	Huntingdon	C	Surveyor-General of Ordnance 1841-6; Major-General 1854; Sec. of State for War 1858.
Hon. Ashley Ponsonby (ys 1st Lord de Mauley)	1850	GG	Ensign	1850	Cirencester	L	DL for Hampshire

Name		Regiment	Rank		Constituency		Notes
Sir J.H.F. Smith (s Major-General Sir F. Sigmund Smith)	1805	RE	Major-General	1851	Chatham	C	Commanding Engineers at Portsmouth
E.A. Somerset (c 7th Duke of Beaufort)	1836	Rifle Brigade	Major	1851	Monmouthshire	C	Equerry to Queen Dowager; Lieutenant-Colonel 1855. Cousin to R.N.F. Kingscote
R.H.R.H. Vyse (2s Colonel Howard Vyse)	1830	RHG	Major	1851	Northamptonshire South	C	Lieutenant-Colonel 1854
Henry Wyndham (2s Earl of Egremont)	1806		Lieutenant-General	1846	Cockermouth	C	General 1854

Source: Compiled from Charles R. Dod, 'The House of Commons in 1853', and contemporary works of reference.

RHG = Royal Horse Guards; LG = Life Guards; CG = Coldstream Guards; SFG = Scots Fusilier Guards; GG = Grenadier Guards; Lt Dr. = Light Dragoons; b = brother; s = son; es = eldest son; 2s = second son; gn = great nephew; n = nephew; ys = younger son; 3s = third son; 5s = fifth son; c = cousin; DL = Deputy-Lieutenant.

family who either sought to interest themselves in politics or who were available to carry on the family tradition. In selecting the family's representative for the
constituency it was irrelevant whether he was a member of
the officer corps or not. Richmond, the proprietary
borough of the Dundas family whose influence in politics
had been recognised in the elevation of Lord Dundas by
the Whigs to an earldom as Earl of Zetland in 1838, was
consistently used to return members of the family to the
Commons. As early as 1768, a Dundas had represented
Richmond, where in the unreformed House of Commons, the
right of electing the member was in the owners of ancient
burgages in the borough, the majority of which were owned
by the family.(10) Selected members of the family followed one another into the Commons without interruption
until 1841. Among these was Lieutenant-General Sir
Robert Dundas (1780-1844). The sixth son of the 1st Lord
Dundas, he had joined the Royal Artillery in 1797 as a
second lieutenant, although he subsequently transferred to
the Royal Engineers with whom he served in the Peninsular
campaigns. A major-general in 1830, Sir Robert was first
elected for Richmond in 1828, joining his nephew Hon.
Thomas Dundas (1795-1873). In the election of 1832,
Thomas was replaced by his younger brother J.C. Dundas
(1806-66), a barrister and a future Lord Lieutenant of the
Orkneys and Shetlands where the family also enjoyed considerable territorial and political influence. In 1835,
both Sir Robert and J.C. Dundas stepped down to be replaced by Thomas again and a nephew, A. Spiers, but Sir
Robert's political career was not yet over. In 1839,
Thomas succeeded as 2nd Earl of Zetland, so Sir Robert was
once more selected to represent the family interest at
Richmond, a task which he filled until the 1841 parliament
when the borough was used to accommodate two ordinary Whig
party men, Colborne and Rich.

A similar pattern of selection arose again in the 1870s,
when in 1872 Lawrence Dundas, the son of J.C. Dundas, was
elected as the twenty-eight-year-old Liberal member. A
lieutenant in the Royal Horse Guards from 1866-71,
Lawrence only served in the Commons for a short period,
for in 1873 he succeeded his uncle as the 3rd Earl, and
his place was taken over by his younger brother who continued to represent the constituency until electoral reform in 1885 merged the borough with the county division.
The 3rd Earl, in the meantime, went on to enjoy a distinguished public career. A Deputy Lieutenant of Stirling-
shire and a magistrate for the North Riding of Yorkshire,
he was appointed a Lord in Waiting to Queen Victoria in
1880, before assuming office as Lord Lieutenant of Ireland
in 1889.

Proprietary boroughs were not only owned by the peerage.
At Eye in Suffolk, a small constituency of 356 electors, a
member of the Kerrison family was returned without a contest for a period of twenty years and six general elections.
From 1824 to 1852, the borough was represented in the
Commons by Lieutenant-General Sir Edward Kerrison, Lord of
the Manor and Recorder of Eye. On his retirement from
political life, he handed the constituency over to his
son Edward Clarence who had married a daughter of the Earl
of Ilchester. A Conservative like his father, Edward
Clarence vowed that he would give 'his Cordial support to
all measures which, without tempering with the welfare of
the rest of the community, may in any way tend to improve
the condition of the agriculturalists', a not unexpected
political philosophy in view of the location of Eye and
the size of the Kerrison estates.
    The peaceful access which the General enjoyed to
Parliament was not however always repeated in other proprietary boroughs, where the post-Reform Act electorate
invited challenges to established family connections. In
one of the most notorious of pocket boroughs, Hertford,
where control by the patron had appeared to be most per-
verse and arbitrary, the power of the proprietors declined
considerably after 1832. In the general election of 1835
the power of the Salisbury monopoly was challenged by
Lord Cowper whose seat, Panshanger, was close to Hertford
and who could appeal to the Whig and Radical element in the
borough. This election brought into the Commons for the
first time the Hon. William F. Cowper (1811-88), brother
of the 6th Earl Cowper who was an Under-Secretary of State
for Foreign Affairs. The nephew of one Prime Minister —
Melbourne — and the step-son of a second — Palmerston —
William Cowper was a Major in the Royal Horse Guards.
Subsequently inheriting Palmerston's estates, Cowper who
was also the brother-in-law of the Earl of Shaftesbury,
the factory reformer, enjoyed a distinguished public
career. Aide-de-camp to the Lord Lieutenant of Ireland,
a Commissioner of Greenwich Hospital, Private Secretary to
Melbourne, and a Lord of the Admiralty in 1853, his later
appointments and the value of his political contribution
were recognized in his elevation to the peerage as Lord
Mount Temple in 1880.(11)
    This control of proprietary boroughs frequently brought
into electoral campaigns military officers, as at Scarborough in 1841 where the candidate was Lieutenant-Colonel
Sir Charles Beaumont Phipps (1801-66), an officer in the
3rd Foot Guards and younger son of the Earl of Mulgrave,
the borough's patron. But the number of these boroughs
after the 1832 Reform Act was relatively small and no more

than fifty-two boroughs at most could be said to be under
the patronage of an individual, although this was supplemented by perhaps a dozen more constituencies characterized
by small electorates, uncontested elections and the return
to Parliament of members of one particular family. While
in these boroughs a number of officers were returned to the
Commons, as in 1853 when they returned sixteen officers,
for other would-be candidates, an alternative means of
access into the Commons was via the corrupt boroughs. The
officer could 'buy' himself a seat by standing in one of
the boroughs which even after 1832 had a reputation for
being rotten to the core or for large-scale bribery of the
electorate. In one of the most notorious of these —
Stafford, two army officers, Captain Gronow and Captain
William Fawkener Chetwynd, were returned in the first
post-Reform Act election. Few officers, however, could
afford the high expense of these elections. A Parliamentary committee of inquiry which sat in 1833 found that in
Stafford the price of votes which had started at £2.10
for a single vote had climbed to £10 before polling day.
One agent alone paid out over £1,000 in bribes to between
400 and 500 voters, and it was calculated that of the 1,000
voters, probably some 850 were bribed.(12) In the next
election of 1835 when the threat of disenfranchisement hung
over the borough, there was no allegation of bribery and
Chetwynd retained his seat, but in the subsequent election
of 1837 when Chetwynd was again returned, it was alleged
that Greville spent £2,500 without success on his candi-
dature in a constituency which he argued had been 'from
time immemorial a corrupt borough'.(13)

   In the parliamentary inquiry of 1836 which for the
second time investigated the charges of corruption in
Stafford, it was alleged that Chetwynd had been elected
not so much through bribery but because of his popularity
as a local candidate.(14) The House of Commons was not
easily convinced by this argument and the high cost of
elections in boroughs such as this and others famed for
their venality, put them out of the reach of most members
of the military establishment. In addition, officers
who came from families who could afford expenses of this
kind were usually able to enter the Commons because of the
territorial influence which their families wielded in the
counties. In Ireland this was particularly noticeable,
and Mervyn Archdall, a Captain in the Inniskilling
Dragoons, faced no electioneering problems when in 1834,
as a twenty-two-year-old, he first sat for the county of
Fermanagh in place of his uncle, General Archdall of
Castle Archdall. Consistently, families such as the Arch-
dalls exercised a control over elections which stemmed

from their property and position as members of the landed
interest. General the Duke of Richmond not only had the
ability to nominate one member for the proprietary borough
of Chichester, where a Lennox was returned from 1832-52,
but his influence extended over a wide area of that part
of Sussex around his seat at Goodwood. The Duke of Gordon
was a dominating figure in Scottish politics from his
territorial position alone. In Cumberland the division of
the county between the predominant influence of the Whig
Earl of Carlisle in the eastern division and the Tory
Earl of Lonsdale in the west, brought into the Commons
military members of their families such as Colonel the
Hon. Henry Cecil Lowther, MP for Westmorland from 1812-68
and his son Henry Lowther, first elected without opposition for West Cumberland in 1847. A Cambridge graduate,
Henry Lowther who served in the 1st Life Guards from 1841
to 1854 represented the constituency until 1872 when he
succeeded as 3rd Earl of Lonsdale.(15)

Nor was this territorial influence limited to the
peerage alone. In Wiltshire, one of the most influential
Tory landowners in the county was Joseph Neeld, an East
India and Bank proprietor, who had inherited a fortune
from a relative, Mr Rundell, a jeweller of Ludgate Hill,
and who through his marriage in 1831 to Lady Caroline
Mary Cooper was connected with both the Shaftesbury and
Marlborough families. Joseph Neeld sat for the borough
of Chippenham from 1832 until his death in 1856, whilst
his younger brother John, who had married Harriet, the
daughter of Major-General Dickson of Blenheim House,
Berks, was the parliamentary member for Cricklade. When
Joseph's sister Mary married Captain Henry George Boldero
of the Royal Engineers, a Parliamentary seat was found for
him on his retirement from the Scientific Corps in 1830 as
the second member for Chippenham. Another conservative,
Boldero who had worked hard to ensure that Chippenham
kept its two seats during the debate on the Reform Bill,
was Clerk of the Ordnance, a junior ministerial appointment, from 1841 to 1846.

Another officer who came into Parliament because he had
married into a family who controlled a proprietary borough
was Major Sir William Payne-Gallwey, MP for Thirsk from
1851-80. At the time of the Reform Act the franchise in
Thirsk belonged to some fifty burgage tenements, forty-
nine of which were owned by Sir Robert Frankland of
Thirkelby, and the enlargement of the franchise in 1832
made little immediate difference to the character of the
constituency. Sir Robert died in 1849 and was succeeded
by five daughters and co-heiresses, one of whom, Emily,
the third daughter, had married Sir William Payne-Gallwey

in 1847. Originally Sir William was returned without
opposition, but an indication of the changes which had
taken place in the English political scene was seen in
the election of 1874 when Sir William defeated his Liberal
rival, Major H.W. Stapylton of the 2nd Dragoon Guards, by
only one vote. Six years later, Sir W.A. Frankland,
grandson of Sir Robert Frankland and a lieutenant-colonel
in the Royal Engineers, could only poll ten votes in the
Thirsk election, the successful candidate, the Hon. L.P.
Dawnay who had retired the previous year as a captain
from the Coldstream Guards, heading the poll with 485
votes.

When this territorial influence was wielded by the traditional military families who had a long record of army
service, this brought into the Commons during the earlier
years of the Victorian period, an exceptional number of
service officers. This was particularly noticeable in the
county constituency of South Staffordshire and in the
Borough of Lichfield where the interest of the two Whig
families of the Ansons and Pagets was most marked. In
1853, the county was represented by General the Hon. George
Anson (1797-1857) and Lord Lewisham, both of whom had been
elected without opposition. The General, at this time
fifty-six years of age, was not lacking in political experience. The brother of the 1st Earl of Lichfield,
Melbourne's Postmaster-General, he had first been elected
in 1818 for Yarmouth. Defeated there in 1835, he then
contested South Staffordshire in 1835 in a by-election
which even during an era of electoral violence, was
renowned for its extreme disorder. The chief feature was
the violence shown towards Tory voters by a mob who were
Anson's supporters, the violence increasing to such an
extent on the second day of polling that a troop of
dragoons had to be moved into the town. For a time the
dragoons remained passive under the attack of the rioters
but eventually the troops were forced to open fire.(16)
Anson was not elected on this occasion, but he took the
seat in 1836, and after a very short spell as member for
Stoke in 1836-7, he became the permanent county representative from 1837 onwards, during which time he occupied
junior ministerial office as Clerk to the Ordnance. At
Lichfield, 'the prize-gem in the jewel case of the Whig
magnates',(17) Viscount Anson, the General's nephew, who
was a precis writer in the Foreign Office and a captain
in the Staffordshire Yeomanry Cavalry, took over the
family seat held from 1832-41 by his great uncle, General
Sir George Anson (1769-1849). His colleague, Lord Alfred
Paget who had been a Lieutenant in the Royal Horse Guards
until he became a Major unattached in 1845, was Chief

Equerry and Clerk Marshal to the Queen, eventually receiving promotion to the rank of General.

The Ansons and the Pagets were among the great families
of English Whiggery and their political involvement and
influence was extended through a series of marriages with
other members of the political and military élite.
General the Hon. George Anson, for example, had married
into the Forester family of Shropshire. One of his
brothers-in-law, Lord Forester, was Captain of the Gentle-
men-at-Arms; another, the Rt. Hon. George Cecil Forester
of Willey Park, Salop, was a former Lieutenant-Colonel in
the Royal Horse Guards. The family borough of Wenlock,
a former close corporation did not change its allegiance
to the Foresters after 1832, and George Forester was returned for the borough at every election from 1832 to
1865. Elizabeth, the General's sister, married the 3rd
Lord Waterpark (1793-1863), an Irish peer and a Cavendish
who represented Lichfield from 1854-6 after Viscount Anson
had succeeded as 2nd Earl of Lichfield. Another sister,
Anna, married the 4th Earl of Rosebery (1783-1868) and
subsequently their grandson was the Liberal Premier in
1894-5. Their daughter Anne married in 1848, as his
third wife, Henry Tufnell the Whig Chief Whip, the cadet
member of an East Anglian landed family who was the son,
grandson and great-grandson of MPs. Tufnell had successively married into the large Whig families of England,
for while his third wife was related to the Ansons through
her mother, his second wife Frances, whom he had married
in 1844, was the half-sister of Captain George Byng,
subsequently 2nd Earl of Strafford. Since Byng when he
was a young army officer had himself married into the
Paget family, the complex family network which linked
General George Anson to Henry Tufnell also linked the latter with Lord Alfred Paget who was the MP for Lichfield.

The Byng family, too, produced officers and politicians.
The Hon. George Byng (1830-98), who was the 2nd Earl of
Stafford's eldest son, was Liberal MP for Tavistock in
1852 and for Middlesex from 1852-74. Subsequently First
Civil Service Commissioner, he filled several junior
ministerial appointments as an Under-Secretary of State
before he succeeded as 3rd Earl in 1886. His younger
brother Sir Henry Byng (1831-99), who became the 4th
Earl, was a Lieutenant-Colonel in the Coldstream Guards
who had been a Page of Honour to Queen Victoria from
1840-7, a Groom-in-Waiting from 1872-4, and an Equerry
from 1874 onwards. Their half-brother Hon. Lionel Byng
(1858-1915), served in the Royal Horse Guards from 1878-
1905, taking part in the Second South African War from
1900-2 with Thorneycroft's Mounted Infantry. The Pagets,

in turn, were an extended family of considerable political
influence, a fact which prompted Melbourne to comment that
they got on very well without education. Four of the
Marquess of Anglesey's sons were Members of Parliament.
Captain Lord William RN (1803-95) was MP for Andover from
1841-7. Admiral Lord Clarence (1811-95) represented
Sandwich, a borough with a long history of Admiralty
influence from 1847-52 and from 1857-66. General Lord
Alfred (1816-88) was the Lichfield member from 1837 to
1865 while General Lord George (1818-80) represented
Beaumaris which was closely associated with a family seat
at Plasnewydd, Anglesey, from 1847-57. Their eldest
brother, Colonel the Earl of Uxbridge (1797-1869), was a
Lord-in-Waiting in 1837 and Lord Chamberlain two years
later. In due course his son represented South Staffordshire in the Commons as successor to General George Anson.

Seen against this background, the military members of
the Anson and Paget families who were active in politics,
were primarily representing a family interest rather than
any commitment to the army. The territorial influence
wielded by the Earl of Lichfield and the Marquess of
Anglesey enabled them to promote the political careers
of members of their respective families, irrespective of
whether they were or were not coincidentally officers.
Neither family, however, was at the very centre of power
politics. Their 'golden' year was probably in 1846 when
three of the Pagets and one of the Ansons, together with
George Byng, were all members of Russell's administration,
but their political influence was consistently recognised
particularly by their critics. In 1852 when Aberdeen was
seeking to form a coalition government, he refused to
follow the Whig policy of conciliation in Ireland, an
action which led Sir James Graham, a leading Peelite, to
argue that 'it was of more consequence to conciliate that
large part of the Empire than to provide for the Ansons
and Pagets'.(18)

The use which was made by these territorial magnates
of their proprietary boroughs to bring into politics the
heir to the title, was also indicative of a third reason
why members of the officer corps became involved in the
House of Commons. When Lawrence Dundas was elected for
Richmond in 1872, the 2nd Earl of Zetland to whom he was
the heir was already seventy-seven years of age. The
short period he spent in the House of Commons before he
succeeded to the title a year later was thus part of a
socialization process designed to equip the future peer
for his eventual responsibilities as a senior member of
the landed interest. In this programme, education at a
public school, a further course at university or Sandhurst
and a period of service in a fashionable regiment was a

prelude to a short spell as a member of the House of Commons. This form of the socialization process was not inevitable, since in many instances it was the second or third son, rather than the heir to the title, who joined the military establishment. Nevertheless, there were a large number of examples of this process in operation during the Victorian period. Viscount Seaham (1821-84) was typical of this group of officers in the sense that he spent three years in the Life Guards and seven years in the House of Commons before succeeding as the 5th Marquess of Londonderry.

In many ways, however, the importance of this socialization process was more clearly seen when it concerned not the members of the aristocracy but aspiring members of the ruling class. These were men, who, coming from the minor landed or professional interest, sought to use the twin avenues of a commission in the military and membership of the House of Commons as a means of individual social advancement. This was clearly brought out in the South Staffordshire election of 1837 when one of the two Conservative candidates who opposed General George Anson and his colleague Sir John Wrottesley was Richard Dyott of Freeford, Staffordshire. Dyott, a young Captain in the army, was the son of General Dyott (1761-1847), a magistrate and Deputy-Lieutenant of the county. Hitherto young Dyott had taken no part in politics and one of the factors which prompted the General to put forward the name of his son was the belief that 'the application of so influential a list of electors would naturally introduce him to the county, and place him in a situation of high respectability in future life'.(19) Dyott was unsuccessful in the election, although he was not bottom of the poll, and the election expenses which he had incurred did not prevent him from seeking a Parliamentary seat again in 1841. The complicated discussions which took place between the Conservatives and the Whigs, and within the Conservative party, showed, however, the problems faced by an aspiring candidate who lacked family influence or who was not a member of the senior sections of the landed interest. When the Conservatives invited Dyott to head the list of candidates to be put up with Viscount Ingestre, son of Earl Talbot, who had been successful in 1837, they expected the Dyotts to contribute £1,000 towards the election expenses. This Dyott declined to do, and he decided instead to contest the borough seat at Lichfield where he hoped to replace Lord Alfred Paget. Again Dyott lost, albeit by a margin of only eight votes, but the real irony was that there were no elections in the two county constituencies of North and South Staffordshire. In the event, Viscount Anson, a Whig, and Ingestre, a Conserva-

tive, were returned for the southern division, an outcome
which infuriated the Dyotts. The General on being told
that a compromise had been decided upon at a Carlton
Club meeting in London, issued a strong protest at any
meeting which presumed to ignore the local conservative
associations. The Dyotts naturally concluded that they
had been 'squeezed out' by the wealthy proprietors of
Staffordshire who, having decided to forego the expenses
of an election, quietly arranged the results in two constituencies, each returning two members, among themselves.
The truth of the matter, the General concluded, was that
the compromise had been decided upon because there was 'a
considerable material which his lordship (Ingestre)
lacked, Money'.(20)

This example shows the difficulties faced by those
officers who, in seeking a parliamentary career, lacked
either family interest, territorial influence or wealth.
Throughout the Victorian period until 1885, election
expenses were very considerable. For the three elections
of 1868, 1874 and 1880, it has been calculated that the
average cost of a county contest was about £3,000; in
borough constituencies it varied from £742 in 1874 to
£1,212 in 1880.(21) Richard Dyott went on to be a
Staffordshire magistrate, lieutenant-colonel of the county
militia and High Sheriff (1857) but, despite his father's
wishes, he was unable to break into the closed circle of
national politics. In contrast, the three sons of the
1st Duke of Abercorn found no such difficulty barring
their way. The eldest son, Marquess of Hamilton (1838-
1913), was the Member of Parliament for the Irish constituency of County Donegal for twenty years from 1860-80
before he succeeded to the title and estates of about
26,000 acres in 1885. The second son, Lord C.J. Hamilton
(b. 1843) served in the Grenadier Guards from 1862-7 and
then successively represented Londonderry, King's Lynn
and Liverpool in the Commons. The third son, who was also
Lansdowne's brother-in-law, first entered Parliament in
1868. Following his successful maiden speech on the Irish
Church Bill, it was decided that Lord George would give
up the army as a career and join his brothers in politics,
a decision which ultimately brought him posts as Under-
Secretary of State for India in 1874-8, First Lord of the
Admiralty, and Secretary for India in Salisbury's third
administration of 1895.(22)

For other officers who lacked this family interest,
their entry into political life consistently depended
during these years on their receiving the benefits of some
form of political patronage, although it is very evident
that such officers formed only a small part of the military

interest in the Commons. The majority of officers who were also Members of Parliament, came almost without exception from families with some form of political interest, and although this may have changed to a limited extent after the electoral reforms of 1885, this relationship was a constant feature of political life before this date.

This dependency on family interest was clearly brought out in the parliamentary representation for the six northern counties of England during the period from 1837 to 1885.(23) Fifty-seven of the successful candidates had been officers in the army, and, of these, thirty were 'aristocratic' in the sense that they were either peers or were the near kin of title holders who retained a territorial influence in these areas. Thus, in 1841, 1,793 electors in the East Riding of Yorkshire signed a requisition urging Lord Hotham not to retire from Parliament but to accept the nomination for the Riding. Lord Hotham (1794-1870), a Major-General in the army who had served in the Peninsula War and at Waterloo with the Coldstream Guards, was an Irish peer. Having inherited the title in 1814, he had sat for Leominster from 1820 onwards, but in 1841 he accepted the invitation to contest the Riding which he represented from 1841-68. A confirmed opponent of the Corn Law repeal movement, Lord Hotham had inherited a considerable estate producing £15,742 a year, and judicious purchase of land in the 1820s increased both the size of the estate and the income it produced, until in 1874, 20,352 acres yielded over £26,000.

In some constituencies, a few families dominated this representation. In Durham, the Marquis of Londonderry and the Earl of Durham continually exercised their patronage to return their nominees to the Commons. When Viscount Seaham gave up his seat at North Durham, he was succeeded by his younger brother, Lord A.D. Tempest, a Lieutenant-Colonel in the Scots Fusilier Guards whose army service from 1843-59 included service in the Crimea. A military Lambton represented Durham in the 1820s and again in the 1880s when Lieutenant Hon. F.W. Lambton, twin son of the 2nd Earl of Durham and a future Earl himself, left the Coldstream Guards in 1880 to sit as Liberal member for South Durham. In Northumberland, the Percy family, from their castle at Alnwick where they lived a life of feudal splendour, consistently dominated the political life of the northern part of the county. Lord Lovaine (1810-99), a former Captain in the Grenadier Guards, represented North Northumberland from 1852-65, and then handed on the seat to his younger brother Lord J.M. Percy, a Crimean VC, whose twenty-six years in the army from 1836-62 culminated

in his promotion to General in 1877. In the North Riding
of Yorkshire, the conservative interest was for years
dominated by the Dawnays (Viscount Downe) and the Dun-
combes (Earl of Feversham), both families providing a
succession of officers and politicians. Nor was this
representation confined to county constituencies. At
Wigan, military Lindsays represented the borough in 1825-
31, 1845-7, 1859-66 and again in 1874-80, when Lord
Lindsay, who like the other members of his family had
served in the Grenadier Guards, represented the constituency until he succeeded as 9th Earl of Crawford and
Balcarres in 1880. Preston, dominated by the family
influence of the Earl of Derby, similarly returned members
of the Stanley family throughout this period, many of whom
had enjoyed a military career before they went into politics.

Family interest was an equally important factor in the
election to the Commons of those officers who were not
members of these aristocratic families. Seven of them in
these northern counties were carrying on a family tradition
of parliamentary membership. These were candidates who inherited a parliamentary seat as much as did their aristocratic colleagues. Lieutenant-General R.J. Feilden
(b. 1824) of Wilton Park, Blackburn, the eldest son of
J. Feilden, MP for Blackburn, followed his father's example
by representing North Lancashire in the House from 1880-5.
At the same time the family military tradition was furthered by the General's son, Cecil Feilden (1863-1902)
who after going through Eton and Sandhurst was commissioned
in the Royal Scots Greys in 1882. A Captain in 1891 and a
Major ten years later, Cecil was ADC to the Lord-Lieutenant
of Ireland from 1891-5 and then Private Secretary to
Wolseley when he was Commander-in-Chief. General Feilden
took over as Conservative member for North Lancashire from
another family with a tradition of military and political
service. From 1844-7 the seat had been held by J.T.
Clifton, High Sheriff of the county in 1835. From 1874-80,
the constituency was represented by his son T.H. Clifton
who like his father had served in the 1st Life Guards. In
the south of the county, one of the parliamentary repre-
sentatives was Captain W.J. Legh (1838-98) of the 21st
Foot, who had entered the army in 1848. The Legh family
were one of the great, if not the greatest, non-aristo-
cratic families of England, for their claim to really
ancient descent was supplemented by their wealth derived
from the possession of 13,800 acres in or around Stockport
in Cheshire, and by their considerable political influence.
A Legh had represented Lancashire in Parliament from 1698
onwards and W.J. Legh, whose political power was recognized

in his elevation to the peerage as Lord Newton in 1892,
followed this family tradition. A similar tradition was
carried on by the Beaumont family in Northumberland and
Durham. Thomas Wentworth Beaumont, MP for Northumberland
in the 1833 and 1835 parliaments, fulfilled completely the
contemporary ideal of what a politician should be, for he
was thoroughly independent, with no party allegiance, great
integrity of character and an income of almost £100,000 a
year.(24) His nephew, Major F.E.B. Beaumont, whose service
in the Royal Engineers from 1852-77 included action in the
Crimea and in the Indian Mutiny, was elected to represent
South Durham as a Liberal from 1868-80, while his son Wentworth Beaumont, created Lord Allendale in 1906, represented
South Northumberland from 1852 to 1885. This influence
could also be important in other ways in promoting the political career of officers. Sometimes it was used to develop the political career of an officer who was not a
relative but who could be relied upon to carry out the
wishes of the family. At Cockermouth in 1867, political influence was used by the family who retained a traditional
hold over the constituency, to bring into the Commons Major
A.G. Thompson whose father was lessee of the fisheries
owned by the Wyndham family. From Cockermouth Castle, the
Wyndhams whose landed interest included estates in Cumberland, East Yorkshire and Wiltshire, shared the political
patronage of West Cumberland with the Lowther family. Normally this was used to promote the parliamentary careers
of members of the family. Thus General Henry Wyndham,
first commissioned in 1806, and son of the Earl of Egre-
mont, represented Cockermouth from 1852-7 and West Cumberland from 1857-60. The latter seat was then held until
1885 by his nephew, Hon. Percy Wyndham (1835-1911) third
son of Lord Leconfield and an officer in the Coldstream
Guards who had retired in 1855 after serving in the Crimea.
The family political tradition was carried on by Percy's
son, Rt Hon. George Wyndham (1863-1913), a distinguished
Victorian statesman. After Eton and Sandhurst, George
Wyndham served with the Coldstream Guards in the Suakim
Campaign and Cyprus 1885, before he began a brilliant parliamentary career as Private Secretary to Rt Hon. A.J.
Balfour. A Cabinet Minister in 1902, and Chief Secretary
for Ireland from 1900-5, George Wyndham's political career
was cut short by his early death in 1913 at the age of fifty.
     In other instances this interest had begun to change
in character, where it was an influence based not on
territorial possessions or traditional political representation, but on an involvement with the party machinery.
As this became more institutionalized, it developed a
patronage of its own so that a strong local association

could select as a candidate someone of their own choice,
unaffected by the wishes of the landed proprietors or the
Carlton Club. At Rochdale in 1857-9, Lieutenant Sir A.
Ramsay of the 85th Foot, eldest son of Sir Alexander Ramsay of Balmain, owed his brief political career not to the
territorial influence of his father but to the position
of his father-in-law as the President of the South Lancashire Conservative Association. Where this patronage
was exercised by the central party machinery its importance was more marked, for it was then used to bring into
political life selected military officers whose presence
was welcomed or needed by the two major political parties.
A proprietary borough such as Ripon which had been placed
at the disposal òf the Conservative Party by the proprie-
tress, Miss Elizabeth Lawrence (1761-1845), was thus
consistently used until her death for this purpose. Among
the party nominees who entered the Commons by this means
was General Sir J.C. d'Albiac, a Peninsular veteran, who
represented the borough from 1835-7, until he was replaced
by Sir Edward Sugden who had been Solicitor-General in
1829 and Lord Chancellor for Ireland in Peel's 1834-5
ministry. After the death of Miss Lawrence, her estates
passed to the Earl de Grey and with a new patron in the
borough, its political affiliations changed. From 1871-4,
Major-General Sir K.H. Storks was MP for Ripon, his
liberal opinions reflecting the radicalism of the 1st
Marquess of Ripon and 3rd Earl de Grey (1827-1909) who as
Viscount Goderich had taken an active part in the affairs
of the 'Amalgamated Engineers' at the time of the 1852
strike, and who had declared himself to 'feel great interest in all that concerns the improvement of the condition
of the working classes'.

This wish on the part of both the political parties to
sponsor the return of officers to the House of Commons was
a fourth reason why members of the Victorian army parti-
cipated in politics. The parties were motivated by two
main considerations. It was recognized that in some constituencies, particularly the dockyard and garrison towns,
a member of the armed forces was potentially a more suitable candidate than a civilian. This did not mean, however, that the parties necessarily selected a nominee from
outside the small group whose links with the landed
interest were the main criteria of their suitability for
membership of the Commons. Party choice was based on an
appreciation that in these constituencies with their small
electorate, the acquisition of a few extra votes by an
acceptable and attractive military candidate could determine the outcome of the election. At Devonport in 1852
where the public establishments of the port gave the

Government considerable influence in the elections, the
Whig Chief Whip, Henry Tufnell and the Conservative
Lieutenant-General Sir George Berkeley were the successful
candidates. Berkeley defeated Sir James Romilly, a former
solicitor-general in Russell's minstry by ten votes in an
election in which the polling was extremely close, and for
the first time in its history the two seats in the borough
were shared between the two political parties. Frequently
both parties put up service candidates. At Chatham which
returned one member to the House, George Byng who had
replaced his Whig colleague Lieutenant-Colonel Maberly as
MP in 1834, was defeated by the Tory candidate Admiral Sir
John Poo Beresford in 1835, but Byng retained the seat in
1837, 1841 and 1847. Conversely, in 1852 Major-General
Sir John Smith of the Royal Engineers, an author and a
former Inspector-General of Railways defeated his Whig
rival, Admiral Sir James Stirling. Captain Lord Charles
Pelham-Clinton was defeated at Sandwich, another 'government' borough, in 1847 although he was returned unopposed
in 1852. At Harwich in the same year, when two Tory candidates defeated the nominees of the Whigs, Captain C.D.
Warburton of the Royal Artillery was bottom of the poll,
but as Gash points out, in Harwich, as in Falmouth and
Penryn, a decline in government influences had been replaced by the power of the highest private bidder,(25)
and ultimately, this change modified the closeness of the
association between these constituencies and their officer
candidates.

A second and more consistent explanation of the willingness of political parties to sponsor members of the military as candidates, was their need to have in the Commons
a number of officers who could fill specific appointments
at junior ministerial level. These were the 'Ordnance'
posts - Surveyor-General, Clerk and Secretary — which
were usually, but not always, filled by military Members
of Parliament. In 1853, when Lord Raglan was Master-
General of the Ordnance, the Surveyor-General was the Hon
Lauderdale Maule (1807-54), a Lieutenant-Colonel in the
79th Highlanders and MP for Forfar, while the Clerk was
an Irish MP, William Monsell of County Limerick who was a
Director of the Limerick and Waterford Railway. In comparison, in 1846 all these posts were held by officers
with the Marquess of Anglesey as Master-General,General
C.R. Fox as Surveyor-General and Colonel Hon. George Anson
as Clerk. The need to have these office holders in the
Commons was, however, challenged by the Northbrook Committee upon whose recommendations the War Office Act of
1870 was based. The Committee made it clear that successful military reorganization depended in no small measure

upon the appointment of suitably qualified individuals to
the post of Surveyor-General of the Ordnance, the holder
of which was to assume in the revised organization control
over all administrative duties in the War Office except
those of pay and finance which were to be controlled by
the Financial Secretary. In commenting on the office of
Surveyor-General, the Committee, however, rejected the
argument that this appointment had to be filled by an
officer who was a politician:(26)

> It would be unfortunate if the appointment came to be
> considered as one which must, as a matter of course, be
> conferred upon a Member of Parliament. It would be
> sufficient, in our opinion, that the office should be
> classed with those of the Naval Members of the Board
> of Admiralty, who form part of the political admini-
> stration of the day, are eligible to sit in the House
> of Commons, but need not necessarily always be Members
> of Parliament.

Although Cardwell accepted the Committe's other recom-
mendations, he did not entirely follow this advice. When
Sir Henry Storks was appointed as Surveyor-General in 1870,
a seat was found for him at Ripon the following year, and
Cardwell continued the long-established tradition that the
Surveyor-General was to be both a politician and a soldier.
Similarly, the nomination of Storks to a 'safe seat' was
comparable with earlier decisions which had brought
Lieutenant-General Sir Henry Fane, when Surveyor-General,
into Parliament as one of the two members for Hastings,
and General Sir Rufus Donkin formerly Whig representative
for Berwick from 1832-7, into the Commons as a member for
Sandwich in 1839. It was evident, however, that with the
abolition of the post of Clerk to the Ordnance in 1855,
the formal participation of the officer corps in decision-
making had been subjected to some limitation. In recognizing this, Cardwell drew attention again to the need for
the Government of the day to have in the Commons among
their supporters, qualified military spokesmen:(27)

> I am painfully conscious of the impossible situation in
> which I am placed as regards military knowledge, and
> consequent powers of dealing with military subjects.
> I have spared no pains to learn all that I could, and
> the more I know the more conscious I am how small a
> proportion it bears to what must be known if the Depart-
> ment is to be properly represented in Parliamentary dis-
> cussion. If there is no soldier in the House of Com-
> mons, who can speak with that sort of knowledge which
> springs from a life spent in the service, the Government
> will come to grief.

Cardwell's novel solution to this problem was to propose that General Sir William Mansfield (later Lord Sandhurst), a former Commander-in-Chief in India, should be appointed to the major political office of Secretary of State for War. It was a suggestion which was rejected out of hand by Gladstone who argued that 'the qualities of a good administrator and statesman went further to make a good War Minister than those of a good soldier'.(28)   The arguments and discussions between Gladstone and Cardwell foreshadowed some of the strains in civil-military relationships which were to develop later in the Victorian period. They also drew attention to the fundamental question of the value of the contribution made by military Members of Parliament to political life. There were officers in the Commons whose activities in day-to-day political life were similar to those of Lord Raglan who during the years when he represented Truro as a Tory did not, at any time, contribute towards the debates in the House. Nor did these members necessarily become any more involved with their constituents. General Jonathan Peel (b. 1799) brother of the Prime Minister, who sat for Huntingdon from 1831 onwards, claimed that in all the years during which he represented the constituency the only question he was asked by the electorate was, 'How are you?'(29)

Some officers took up their Commons responsibilities with reluctance. Percy Wyndham accepted parliamentary work 'as a duty like other duties but rather an irksome one. During the 25 years in which he sat for West Cumberland, he won great esteem as a private member and did valuable work on many committees. But he held his own principles unswervingly and they were not always those of his party.'(30) On the other hand, while only one officer became a Prime Minister, many others went on to enjoy a distinguished political career in junior or ministerial appointments in and out of the Cabinet. Of those officers with a lengthy service career, Sir George Murray, second son of Sir William Murray of Ochtertyre, was a typical example of the contribution which an officer could make to political life. Wellington's Quartermaster-General in the Peninsula, Sir George was subsequently MP for Perthshire and, coincidentally, Governor of the Royal Military College at Woolwich. Taking up an active military appointment as Commander-in-Chief in Ireland, Sir George resigned this in 1828 on being appointed Secretary of State for the Colonies and War, and in Victoria's reign he continued to sit in the Commons as member for Manchester.

Officers who came from the landed interest and whose military career was but a part of their socialization

process, enjoyed careers comparable with those enjoyed by
other members of this ruling class. Captain Hon. F.A.
Stanley (1841-1908), later Lord Stanley of Preston and
16th Earl of Derby, spent seven years in the Grenadier
Guards from 1858-65 when he was elected to represent
Preston in the Commons. The son of the 14th Earl who was
Prime Minister on three occasions between 1852 and 1868,
and brother of the 15th Earl who was Foreign Secretary in
1866-8 and 1874-8, F.A. Stanley enjoyed a distinguished
political career which included posts as Secretary of
State for War (1878-80), for the Colonies (1885-6),
President of the Board of Trade (1886-8) and Governor-
General of Canada (1888-93). A similar career pattern in
which political appointments at home were combined with
public appointments in the Colonies, was enjoyed by Sir
James Fergusson of Kilkerran (1832-1907) who when a young
Guards officer in the Crimea had been elected in his absence MP for Ayrshire in place of Lieutenant Colonel James
Hunter Blair (Scots Fusilier Guards) who had been killed
at Inkerman. Under Secretary of State for India in 1866-7
and for the Home Department in 1867-8, Fergusson was
successively Governor of South Australia, of New Zealand
and of Bombay before, on returning to Britain in 1885,
he resumed a parliamentary career as MP for Manchester.
In the second phase of his political life, Sir James subsequently became Under-Secretary of State for Foreign
Affairs and finally Postmaster-General in 1891-2.

More rarely, a political and public service career was
combined with senior military appointments, and of these
officers, Field Marshal Viscount Hardinge (1785-1856) was
one of the more notable examples. Secretary-at-War in 1830
and 1841-4, Hardinge, who had represented Durham City in
the Commons from 1820-30 and Launceston from 1837 onwards,
became Governor-General of India in 1844, subsequently
waiving his right to command in the field so that he could
serve as deputy to General Sir Hugh Gough in the First
Sikh War. He left India in 1847 with a very high reputation, and five years later he succeeded the Duke of Wellington as Commander-in-Chief.

For other officers, their political contribution and
their participation in decision-making was less clearly
linked to active involvement in the Commons or Lords.
Some, in their capacity as the heads of families with considerable political patronage, used this power to influence
constituency results, so that the families continued to be
an important factor in party and electoral calculations.
For a few officers, their interest in the cut and thrust
of party politics brought them positions of importance in
the party machinery. General the Earl of Rosslyn was one

of the most influential of the Tory party's managers in
the 1830s, and a leading member of the party election committee. One of the principal executors of the electioneering fund which was used to assist selected candidates
in the constituencies, Rosslyn, together with Hardinge,
was a member of the committee which managed the Carlton
Club, a point of union and the centre of organization for
the whole of the party. The power and patronage which
these positions gave to Rosslyn were very considerable and
far-reaching. In 1835, he was advising Peel on the pro-
priety of subsidizing Disraeli's election attempt in
Buckinghamshire. In the same year, 'he included among his
other activities in London a visit to the Admiralty to
obtain votes for various conservative candidates'.(31)
Rosslyn's associate in this control of the party machinery
was another former officer, William Holmes, Treasurer of
the Ordnance (1820-30) and the Tory Chief Whip in the unreformed House of Commons. The party's principal election
manager until he was replaced by F.R. Bonham in 1832,
Holmes was a former captain in the West India Regiment.
    Rosslyn and Holmes were atypical members of the officer
corps in the way in which they became involved in the
management of party politics, and it was perhaps significant that they were at the height of their political power
in the unreformed House of Commons. Indeed, it can be
argued that all officers who became involved in national
politics as member of the Commons or Lords were unrepresentative of the officer corps. Statistically, no more
than a very small percentage of the officer corps could at
any one time be active in Parliament. Even if every peer
and every member of the lower House had been drawn from the
army, this could not have involved at any one time more
than a fifth of all officers. But the outstanding feature
for most of the Victorian period, a characteristic which
perhaps above all others was indicative of the political
attitudes of the army, was the way in which the interpenetration of the officer corps and the ruling class ensured
that the army saw itself to be an integral part of the
parent society. This link did not depend solely for its
effectiveness on the active participation of a small number of officers in the actual political process. It was a
much more complex relationship involving family ties,
shared educational experience and common membership of a
hierarchical social structure. This guaranteed that the
political attitudes of the military for most of the nineteenth century were based on their position in society as
part of a ruling power group. When access to this group
was seen to be 'open', when officers could equate their
interests with those of the power élite, then the army

felt in no way isolated from the remainder of society. It
saw itself as a section of the greater social system sub-
scribing to an élitist ethos which encouraged the maintenance and preservation of a stable political structure.

In part, these links depended for their effectiveness
on the status of the military as part of the landed interest. To a considerable extent, they were founded on the
interpenetration of the political and military élites, not
only in the sense that individual officers had become a
part of that élite, but also in the way in which the kin
of political leaders were to be found throughout the army.
But underlying the whole structure, forming the base on
which the position of the military establishment as a part
of the ruling class had been developed, was the involvement
of the officer corps in the minutiae of local politics and
administration. This area was, for the great majority of
the officer corps who could not be a part of the national
parliament, the means which prevented the military from
becoming a self-contained and relatively isolated social
system. It was an area of considerable potential, providing innumerable appointments which involved officers
both in decision-making and in the execution of policy.
While it lacked the glamour of national politics, it was
the backbone of Victorian public administration, and the
occupancy of appointments within its structure closely involved individual officers in the affairs of local society.

In general terms, these appointments were the posts of
local administration and government in the counties, a
characteristic which strengthened the identification of the
army with a rural élite. In many of the cities and
boroughs, the post-1834 reformed corporations tended to be
drawn from the ranks of an urban middle class who formed a
professional and commercial interest which in the early
part of the Victorian period was not fully represented in
county and national politics. There were exceptions,
particularly where a 'new' town had been developed by a
landed proprietor as at Hartlepool where Lieutenant-
Colonel Lord W.J.F. Powlett (68th Foot), nephew of Major-
General the Duke of Cleveland, was Mayor, but the dominance
of an upper ruling class was primarily of importance in the
rural areas. The reason for this is made clear by
Guttsman:(32)

> The larger landed estate on the other hand was to its
> owners a source of income as well as the focal point
> for a series of activities and public positions. They
> were not only patrons of livings and Justices of the
> Peace but on them rested generally the major share of
> responsibility for the running of village schools, in-
> firmaries, and other local welfare institutions.

For the military officer who was a senior and
influential member of the landed interest, all positions
within the county oligarchy were open to him. In 1853,
nine of Her Majesty's Lords-Lieutenant were drawn from the
officer corps, their territorial responsibilities ranging
from Anglesey (Field Marshal the Marquess of Anglesey) to
Durham (General the Marquis of Londonderry). Nor was this
an atypical year. In 1898, thirteen English counties from
a total of forty-three had as their Lord-Lieutenant a peer
who had served in the army. Of these, some had spent but
a short time in the army, their period in the Victorian
military establishment forming part of the socialization
process which prepared them for this ultimate local responsibility. But the short military career of the Duke of
Northumberland or the Earl of Derby who was Lord Lieutenant
of Lancashire from 1897, could be compared with that of the
8th Duke of Beaufort (1824-99), the Lord Lieutenant of
Monmouthshire. Commissioned in the 1st Life Guards in 1841
after leaving Eton, the Duke had combined a military career
which included appointments as ADC to the Duke of Wellington and Lord Hardinge with the parliamentary representation
of East Gloucestershire from 1846-53. On his succession as
8th Duke in 1853, he combined the management of his 52,000
acres and Court appointments as Master of the Horse with
his military career, until he retired from the army in
1861. The Duke's brother-in-law, the 3rd Earl Howe (1822-
1900), a General who had seen service in the Kaffir War
and the Indian Mutiny during a career which began in the
Grenadier Guards in 1838 and ended in 1881, five years
after he had succeeded his brother as 3rd Earl, was Lord
Lieutenant of Leicestershire.(33) Other officers who were
Lords-Lieutenant had similarly enjoyed a distinguished
political career which prepared them for their subsequent
responsibilities and which complemented the administrative
expertise acquired through military service.
    Despite variations in the length of their military
service or in the extent to which they had been involved
in national politics, the common attitudes of these officers towards their local obligations were essentially
derived from their dual membership of the landed interest
and the officer corps. This was not a prerogative of the
peerage, alone, for other major landowners who were Lords-
Lieutenant, such as Sir Francis Macnaughton (1828-1911)
head of the Clan Macnaughton, a former Lieutenant-Colonel
of the 8th Hussars who was the Lord Lieutenant of County
Antrim, exhibited a similar awareness of their duties. A
positive emphasis was placed upon 'local' identification
and this was a fundamental characteristic of the involvement of these officers in the affairs of the county. It

was not restricted to members of those families who had a
traditional and lengthy association with a community, for
newcomers into an area such as Colonel Lord Wantage in
Berkshire were quick to adapt to the demands of the local
environment. They assumed without difficulty the responsibilities which were inherent in their role as Lord-
Lieutenant of the county, taking an active part in two
main areas of activity. In the first of these, their army
experience was a positive asset, for up to 1870 the Lord-
Lieutenant in each county retained specific military
powers. He appointed all officers to their commissions in
the local militia and, more importantly, he took command
in any emergency of all the local forces of the county in
its fullest geographical sense, including, in the word of
the commission which he held, 'all cities, boroughs,
liberties, places incorporated and privileged and other
places whatsoever within the said county and the limits
and precincts of the same'.(34) The potential importance
of this responsibility varied over the years. During much
of the nineteenth century, the absence of any external
threat to British security suggested that the military
responsibility of the Lord-Lieutenant was no longer an
important part of his function. Yet it was to these heads
of the counties that the War Office issued instructions
when the development of a hostile spirit in France in
1858-9 prompted the re-creation of the Volunteer Force on
a new basis:(35)

> I am directed by the Secretary for War . . . to acquaint
> you that a circular had been issued from this Department,
> informing the Lords Lieutenant of the Counties of the
> conditions on which Her Majesty's Government will recom-
> mend to Her Majesty, the adoption of the services of
> volunteer corps in their respective counties.

Prior to this, some Lords-Lieutenant responding to a
public demand for the resumption of volunteer soldiering,
had already formed various corps in readiness for the time
when their actions would receive official sanction and
support. When this came, in the guise of War Office
Circulars dated 12th and 14th May 1859, all Lords-Lieutenant reacted by organizing and raising volunteer corps
in their respective counties, and by nominating officers
to their commissions. Nor did the assumption by the War
Office in 1871 of responsibility for these volunteer
forces sever the links between the Lieutenancy and the
Volunteers. Individual Lords-Lieutenant continued to
serve as Colonels of the newly-formed regiments, while
this honorific responsibility was often translated into a
more practical form by their families. Sometimes the local
regiment, particularly if it were a yeomanry unit,

appeared to be the private army of a particular territorial
magnate or magnates. Continually in the nineteenth century, the Earl of Chester's Regiment of Yeomanry Cavalry,
to take but one example, appeared to be dominated by two
county families. The Colonel of the regiment in 1898-9
was Piers Egerton-Warburton (1839-1914) of Arley Hall,
Cheshire, a county justice and Conservative Member of
Parliament for mid-Cheshire from 1876-85. The two principal squadrons named after Tatton Hall, the seat of Lord
Egerton, and Eaton Hall, the seat of the Duke of Westminster, were commanded respectively by a brother of Lord
Egerton, Lord Lieutenant of Cheshire from 1868-83, and by
the second son of the Duke who was the Lord Lieutenant from
1883-98. One of the majors in the regiment was Lord
Egerton's eldest son, Wilbraham (1832-1909), who succeeded
him as 2nd Baron in 1883, and who was created Earl Egerton
in 1897.(36)

The Volunteer movement, irrespective of its abilities
as a fighting force, was an important social catalyst
within the county, for it perpetuated the identification
of the army with a rural élite. It strengthened the links
between the landed interest and the military, for it
brought into these regiments at all rank levels, members
of county society. 'Take your gamekeepers as your comrades and any of your labourers that will enrol themselves,' wrote Sir Charles Napier in 'The Times', 'A
gentleman will find no braver or better comrades than
amongst his own immediate neighbours and tenants.'(37)
Through their commissions in these Volunteer forces, half
pay and retired army officers were also linked more closely
with the rural élite. Reviews, field days and annual camps
in the parklands of the landed interest, in common with
hunting and other field sports, generated a social inti-
macy which combined the exclusiveness of the officers'
mess with the prestige of county society. In addition,
when a militia commission was the 'back-door' into the
regular army, in that the young aspirant did not have to
go through Sandhurst or sit the direct entry examination,
many other young officers among whom were future military
leaders such as Wilson and Trenchard, began their military
career in this atmosphere. They were trained in regiments
in which soldiering and the acquisition of military skill
seemed to be subordinate in importance to the need for
officers to acquire a common standard of social behaviour.
They underwent at an impressionable age a socialization
process which considerably affected them. Not all commanding officers adopted the neo-feudal attitudes demonstrated by Colonel Piers Egerton-Warburton who gave one
of his tenants notice to quit in 1893, because he refused

to accept the condition that all the tenant farmers on the
Colonel's estates should be members of the Church of
England and do personal service in the yeomanry cavalry.
(38)   But most senior officers emphasized the need for
their juniors to accept the norms of the county society
and to look upon service in the county's own regiments as
part of their contribution to the maintenance of a particular rural way of life.   In turn, these junior officers
and, indeed, the former regular officers who were commissioned in these volunteer forces, accepted the importance
of the landed interest — officer corps links, and they
adopted attitudes which mirrored those of the rural élite.

In many ways, the activities of these forces endorsed
again the importance in the Victorian army of the guardian
concept, for their ostensible function was to protect the
county against foreign invasion.   Another part of the
guardian role was also implemented in the second area of
activity in which Lords-Lieutenant were involved, that is
in the maintenance of law and order.   Primarily this was
associated with the position of the magistrate in county
society.   The prestige of local politics before the creation of County Councils in 1888 rested upon the right of
this magistracy to be consulted in those wider issues of
policy which went beyond the simple function of the magistrate as the dispenser of local justice.   Magistrates were
very much concerned with any attempt to promote a local
Act of Parliament.   As ex-officio members of the various
statutory authorities established during the Victorian
period to administer such diverse services as the Board of
Guardians set up under the Poor`Law Amendment Act, or the
Highway Boards, they played an integral part in local
politics and administration.   Their role in the local
power structure was clearly summed up by Edmund Burke in
his comments on the life of the Whig statesman, W.
Dowdeswell:(39)

Immersed in the greatest affairs, he never lost the
ancient native generous English character of a country
gentleman.   Disdaining and neglecting no office in life,
he was an ancient municipal magistrate;   with great
care and clear judgement administering justice, maintaining the police, relieving the distress and regulating the manners of the people in his neighbourhood.

At the apex of this power structure stood the Lord
Lieutenant of his county in his capacity as Custos
Rostulorum, the 'keeper of the records'.   His office was
supreme in local society, and, until 1910, appointments
to the magistracy were almost exclusively through the Lord
Lieutenant.   It was, indeed, so unusual for any person to
be added to the Commission except on his recommendation,

that would-be justices were referred to him as a matter of course. Brougham brought this out clearly in a speech in the House of Commons:(40)

Such a thing is hardly ever known as any interference with respect to these nominations by the Lord Chancellor. He looks to the Lord-Lieutenant or rather to the Custos Rostulorum, for the names of proper persons. The Lord-Lieutenant, therefore, as Custos Rostulorum appoints all the Justices of the Peace in his county at his sole will and pleasure.

In dispensing patronage, the Lord-Lieutenant selected these 'proper' persons from a narrow circle of the county population. Initially a property qualification limited the area of recruitment,(41) but even if individuals satisfied the statutory requirement, this, in itself, was not a sufficient reason for their selection. The qualities looked for in a potential Justice of the Peace were based on ascriptive criteria which emphasized the need for candidates to have a position of respect in local society. These were, above all, the qualities of a 'gentleman', that is, a life of leisure which enabled the magistrates to find sufficient time to attend at Petty and Quarter Sessions, an attitude of mind based on a willingness to accept the customs and traditions of rural society and a sense of identification with the ethos of the rural élite. Whig critics of the system of selection were quick to point out that it was 'one of the last remnants of class legislation, vicious in principle and obstructive in operation, by which the administration of justice was subservient to the social elevation of a class'.(42) But while the system persisted, former members of the army were ideal candidates for office. Since there was a close similarity between the qualities looked for in magistrates by the Lord-Lieutenant and those sought from officers by the Commander-in-Chief, the possession of a military commission was a guarantee of social fitness and acceptability. The automatic identification of the officer corps with the landed interest ensured that the military candidates for the magistracy were not tainted by any association with trade, commerce or industry, activities which prejudiced the aspirants' chance of selection.

For many officers, the pattern of nomination resembled the system of patronage which gave them almost inevitable membership of the House of Commons. Younger sons could expect an appointment to the magistracy in the same way that they expected to be selected to represent the family constituency in the Commons. Other officers combined their service as a magistrate with an involvement in local politics which, in the case of Colonel J.M. Clayhills who

had joined the army in 1852 and had served in the Crimea
and Indian Mutiny, was unsuccessful in 1886 when he stood
as Liberal candidate for the Whitby division of North
Yorkshire. Many of these military magistrates were the
cadet members of aristocratic families, their own landed
estates confirming their suitability for appointment and
thus increasing the impression that the County bench was
a closed, family-based élitist institution. In Warwick-
shire, both Lieutenant-General Sir Edward Newdigate-
Newdegate (1852-1908) and his brother Lieutenant-General
Sir Henry Newdigate (1832-1908), for example, were justices. Grandsons of the 3rd Earl of Dartmouth, both
generals had enjoyed a distinguished military career
which, in common with many of their contemporaries, included service in the Crimea and Indian Mutiny, before
they retired to manage their country estates. For the
older brother, who had inherited the Arbury and Harefield
estates from the Rt Hon. C.N. Newdegate, his acreage of
6,900 acres placed him in Bateman's category of the
'Greater Gentry', and this section of the landed interest
was a consistent source of recruitment to the magistracy.
But not all officers could claim to own as large an
estate as the 84,000 Argyllshire acres occupied by Sir
John Campbell-Orde (1827-97), a retired Captain of the
42nd Highland Foot. Many county justices were the military squires. These included the Victorian heroes such
as Lieutenant-General Sir Havelock-Allan (1830-97), a
county justice in Durham and a Deputy-Lieutenant who had
won the VC at Lucknow, Major-General Sir Edward Blackett
(1831-1909) of Maften Hall, Corbridge, a Northumberland
magistrate who had been severely wounded at the Redan in
the Crimean War, and Major-General T. de Courcy Hamilton,
VC (1825-1908) a justice in Gloucestershire.
    But not all officers were so closely connected with the
landed interest, and their appointment to the magistracy
was a concomitant of their post-military career. The
Major of the Tower of London, Lieutenant-General Sir George
Milman (1822-1915) was a justice in both London and Middlesex. At Canterbury, the High Seneschal of the Cathedral,
Lieutenant-Colonel Edward Dickenson who had retired from
the 20th Foot, was a county justice who was very active
in county and diocesan business as Commissioner of Sewers,
of Income Tax and Land Tax. In carrying out these administrative functions, the Lieutenant-Colonel was able to
call upon the experience which he had gained within the
Victorian army, and this experience was a valuable asset
to many magistrates. In an era when the county justices
were still very much involved in association with the
army and yeomanry in the maintenance of internal law and

order, the military experience of these magistrates were
particularly welcome. Not all of them could rival the
previous involvement of Sir Digby Mackworth in the suppression of agrarian industrial unrest. A Peninsula and
Waterloo veteran, Mackworth as a Major in the 13th Light
Dragoons had been responsible for suppressing disturbances
in the Forest of Dean and Bristol in 1830 and 1831, services which were rewarded by his creation as a Knight of
the Hanoverian Guelphic Order. In the 1840s as a justice,
Deputy-Lieutenant and High Sheriff of Monmouthshire (1843),
Sir Digby continued to be involved in the maintenance of
law and order. But this was not the only area of activity
in which a military magistrate found his previous experiences to be advantageous. In a period when few other institutions could rival the position of the military as a
complex bureaucratic organization, military administrative
experience was a **positive** asset to the potential magistrate. Indeed some of these officers seemed to dominate
the administrative functions which were carried out by
magistrates at this time. Colonel Sir Robert Gunter (1831-
1905) was not only a magistrate in the West Riding of
Yorkshire, but was also Chairman of Wetherby Petty Sessions,
Chairman of Wetherby Board of Guardians which was responsible for the local administration of the poor law, Chairman of Wetherby District Council and Colonel-Commandant
of the 3rd Prince of Wales Regiment of Militia.

    The steady replacement in the counties of amateur ad-
ministration by full-time professionals partly increase
the participation of officers in local affairs. An extension of the power of the magistracy was particularly
noticeable after the establishment of County police forces
in the 1840s and 1850s to replace the older inefficient
mix of village constables, voluntary associations for ap-
prehending felons and yeomanry forces. Many of the Chief
Constables who took control of these police forces were
drawn from the officer corps. In the Lothians and Peeble-
shire, the Chief Constable, Lieutenant-Colonel Alexander
Borthwick (1839-1914) of the Rifle Brigade, was typical of
many more, but this pattern of appointment was not limited
to the constabulary in rural areas. The Commissioner of
the Metropolitan Police from 1890-1903 was Colonel Sir
Edward Bradford (1836-1911), son of the Rev. W.M.K. Bradford and father of Lieutenant-Colonel Sir Evelyn Bradford
(1896-1914) of the Seaforth Highlanders. The appointment of
these officers as Chief Constable was an important factor
in confirming the hold of the ruling class over the forces
of law and order, for the financial control of the magistracy over the police through quarter sessions was complemented by the way in which the chief constable was drawn

from the same social background: 'Out of his (John
Barker) thirteen children, two sons became Deputy Lieutenants and J.P.s in Suffolk, one served in the army and
crowned an army career by becoming Chief Constable of
Birkenhead.'(43)

   But while the landed interest had successfuly resisted
all attempts in the opening years of the Victorian period
to establish elected county authorities, the growing professionalization of county administration was systematic
of a spirit of change.   Behind a facade of stability, the
functions of Quarter Sessions had begun to alter after
1830, and the establishment of completely elected authorities such as the School Boards and the Sanitary Districts
foreshadowed the end of the patronage system.   The significant result of the 1888 Act which established County
Councils was that it opened a wide area of local administration to the processes of democratic election.   Something of the 'old deference' which had hitherto characterised county administration and life disappeared
following the County Council Act of 1888, even though the
foundations of the social system remained the same.(44)
This placed certain limitations on the traditional control
of county affairs, and the military, in view of its association with the landed interest, was equally affected by
the restrictions which were placed on the power within
the county of the territorial magnates.   In some counties,
particularly those with urban associations, the land-
owning families who had habitually dominated local politics and administration chose not to participate in county
council elections.   In Cheshire, for example, the group
who contested the newly-created council and who came
forward as councillors were mainly the manufacturing and
mercantile families who by 1888 had already secured a place
among the county gentry and magistracy.   The traditional
landed interest preferred to remain apart from the contest,
and only thirty-four county magistrates offered themselves
for election to the County Council.   Among those who were
elected were the older justices such as General Sir
Richard Wilbraham (1811-1900), son-in-law of William
Egerton, a seventy-eight-year-old Crimean veteran who
served on the council until 1892.   Many others either retired after a few years of office as did Colonel Cornwall-
Legh or died within a few years as did Colonel John
Kennedy.(45)

   In other counties, officers came forward more readily
to assume appointments on the new County Councils.   In
Kent, one of the county aldermen was Colonel T.J. Holland
(1836-1910), the historian of the Abyssinian campaign, who
from Mount Ephraim House, Tunbridge Wells, was a Deputy-

Lieutenant and Justice in Kent, and a Justice in both
Surrey and Cornwall. In many rural areas, the election of
a member of the landed interest as the Chairman of the
newly established councils suggested the continued domination of these bodies by the traditional élite. In the
East Riding of Yorkshire, the Chairman of the County
Council was Sir Charles Legard (1846-1901), a former
officer in the 43rd Light Infantry who was the epitome
of the English country gentleman. A champion shot, race-
horse owner and cricketer who was President of the MCC,
Sir Charles played an important part in national politics
as MP for Scarborough (1874-80) and in local administration,
where he was Chairman of the District Council and of the
Board of Guardians. As part of his rural interests he
kept a pack of harriers and a pack of otter hounds, but
he was perhaps best known in the county for his devotion
to voluntary work.

The Chairman of the Somerset Quarter Sessions, the Rt
Hon. Sir Richard Paget (1833-1908), formerly of the 56th
Foot, became the Chairman of the County Council. In the
Isle of Wight, the first Chairman of the County Council
was Lieutenant-General Sir Somerset Gough-Calthorpe, 7th
Baron Calthorpe (1831-1912), who had been an ADC to Raglan
in the Crimea. The Convenor and Deputy Lieutenant of
Berwick, Chairman of the County Council and the Road
Board, was Sir George Houstoun-Boswell (1847-1908) a
former captain in the Grenadier Guards and owner of 5,400
acres. But the essential feature of county life after
1888 was that these members of the landed interest could
no longer expect, as a matter of course, that they would
dominate local administration through their membership of
the magistracy. Some were prepared to stand for election,
but other officers agreed with Lord Caernarvon and Lord
Rosebery that the establishment of County Councils in
1888 was a revolutionary measure which meant 'the dethrone-
ment of the squirearchy'.(46)

The changes which took place in the counties in the
last decade of the Victorian period had a considerable
effect on the attitude of the military to politics. Many
members of the officer corps found for the first time that
their hitherto undisputed right to assume the role of the
governor in county administration was subject to challenge.
Many of them agreed with the eulogistic sentiments expressed by Maitland in 1888, in his commentary on a
'critical situation':(47)

> The hope of securing able and just administration must
> now lie, not in the creation of fancy franchises which
> at best are fleeting, rickety things, but in the charac-
> ter of the work . . . . If possible, men of the same

stamp as those who have hitherto been active at
Quarter Sessions should be obtained; but no tinkering
of the electoral machinery can assure this result.

Nevertheless, groups who had previously been excluded
from participation in local political activity were now
coming forward to contend with the traditional élite for
a share in the distribution of power. And in this area of
activity, the military were subject to the same constraints
as were the whole of the landed interest. Neither had
any particular advantage in the face of a professional
bureaucratic administration. No longer could army officers assume that a right to rule was theirs by divine
right. If they were prepared to compete with other groups,
then their advantages of experience, of popular appeal and
of traditional loyalties might be in their favour, but
their domination of local politics and administration
could no·longer be taken for granted.

Increasingly, therefore, in the last decade of the
nineteenth century, many members of the officer corps
concluded that they were being deliberately excluded from
participation in politics. Certainly some of the actions
taken by the civil power at national level appeared to
confirm this conclusion. This was particularly noticeable
when, in 1880, Childers, the Secretary of State for War,
referred to the Speaker the question of officers, who
were serving on full pay, concomitantly standing as Parliamentary candidates.(48) Speaker Brand, in his reply,
produced an answer which appeared to be a direct attack
on the traditional relationship between the military and
the Commons:(49)

How is a man to serve the Crown and the People at the
same time? In former days when the service of the
people involved little personal attendance in the House,
the two services might be combined in one person; but,
in these days, faithful service of the Crown and People
is almost impossible, except under particular circumstances.

This reply did not immediately prevent an officer continuing to be both a soldier and a politician, and among
the last of the Victorian officers to exercise this dual
role were Captain Sir Herbert Naylor-Leyland (1864-99)
and the Marquess of Hamilton (1869-1913). An officer in
the 2nd Life Guards, Sir Herbert was retired from the
army and resigned his seat in the House. Three years
later, he re-entered the Commons as a 'Liberal Home Ruler',
when he was elected to represent South-West Lancashire. The
Marquess of Hamilton, son of the Duke of Abercorn, was an
officer in the 1st Life Guards from 1892 until he resigned
in 1903. In 1900 he was elected Conservative MP for the

City of Londonderry and from 1903-5, he was Treasurer of
HM's Household. It was, however, significant that a
member of the military élite who was also a Member of
Parliament was, during the 1890s, forced to resign his
seat on appointment to an active senior army appointment.
In 1898, Major-General Hugh McCalmont, a member of the
Wolseley 'ring' and Conservative member for North Antrim,
left the House of Commons when he was appointed General
Commanding, Cork District. One effect of this was to
accentuate a feeling within the officer corps that there
was a growing divergence between the military and the
civil power, in which the former were beginning to play
an ever diminishing part in political life. Officers had
good grounds on which to base their conclusions. From
the 1850s onwards, it was evident that the military were
far from being the single largest occupational group in
the House of Commons. In the Parliament of 1853 when
the total strength of the military interest in the Commons, including officers who had left the army and those
on the retired or half-pay list, was seventy-one, 107
Members were either solicitors or had been called to the
bar. In addition, over one hundred members had more than
a tenuous connection with the commercial and industrial
interest. The representation of the army in the House
of Commons continued to decline in subsequent years,(50)
until by 1898, as Table 20 demonstrates, the military
interest was less than one quarter the size of the legal
interest and one-sixth the size of the commercial and
manufacturing interest.

    Nor could it be assumed that this military interest was
a singularly oriented, cohesive, homogeneous group. There
were, indeed, backbenchers who seemed to be the very
caricature of the military politician. Colonel Sir
Frederick W.J. Fitzwygram (1823-1904), first elected to
represent South Hampshire in 1885, was the epitome of the
traditional ties between the military, the landed interest and politics. Educated at Eton, the Colonel who had
served with the Inniskilling Dragoons in the Crimea, subsequently commanded the Cavalry Brigade at Aldershot,
before taking up an appointment as Inspector-General of
Cavalry. Owning some 2,000 acres of land around his seat
at Leigh Park, Havant, and a further 1,800 acres in Lincolnshire, Sir Frederick, a member of the Carlton Club,
was a perfect squire whose interests were reflected in his
publications, 'Horse and Stables', 'Notes on Shoeing',
'Utilization of Cottage Sewage' and 'Parochial Life Incum-
bencies'. But other officers were more closely associated
with some of the other interests in the House of Commons,
and were more identified with the parent society. In the

TABLE 20   Occupation and political affiliation of MPs:
1898

	Conser-vative	Liberal Unionist	Gladstonian Liberals	Others	Total
Landed interest	91	8	16	3	118
Army	34	2	4	1	41
Navy and other services	19	1	1	2	23
Lawyers	83	14	50	18	165
Other pro-fessionals	4	1	10	5	19
Arts	10	3	14	15	42
Trade and commerce	89	35	86	33	243
Unclassi-fied	7	1	6	5	19
Total	337	64	187	82	670

Source: 'Constitutional Year Book', 1899

same way in which the peerage became involved in commercial and industrial undertakings, so did these military
squires participate more and more in activities which
differed from those of the traditional landed interest.
Sir John Heron-Maxwell (1836-1910), head of the Maxwells
of Poloc and chief of the Clydesdale Maxwells, a Deputy-
Lieutenant and magistrate in the family county of Dumfries,
and a magistrate in Cumberland, Surrey and London, retired
from the 15th King's Hussars in 1865 'to engage in commercial pursuits'. In these he was very successful,
eventually serving as Master of the Merchant Taylors
Company in the City of London.
    The commercial activities of Colonel Gerard Smith
(b. 1838) were another good example of the way in which
officers who were integrated with society could play an
important part in economic development. Member of Parliament for Wycombe from 1882-5 and an unsuccessful candidate in the Holderness Division of East Yorkshire in 1885,
Gerard Smith was the third son of M.T. Smith, MP for
Wycombe from 1847-65 and an East India Company Director,

and grandson of the London banker John Smith of Blendon
Hall, Kent. Commissioned in the Scots Fusilier Guards
in 1857, Colonel Smith, on his retirement in 1874, became
a partner in an old-established Hull banking house. Subsequently a director of the Yorkshire Insurance Co. and a
director of the Hull Dock Co., Smith became the Chairman
of the Hull, Barnsley and West Riding Junction Railway and
Dock Company, a venture which in building a new railway
line from the South Yorkshire coalfield to Hull in 1885
and in opening the Alexandra Dock in the port, owed its
success to his efforts. Gerard Smith was an atypical
officer who preferred to continue a family tradition of
commercial activity rather than identify himself exclusively with the landed interest, but other officers too,
such as A.J. Otway (1822-1912), the Deputy Speaker of the
Commons, who had left the army to read for the Bar, broke
away from the stereotyped pattern which had traditionally
characterised the identification of the military with the
landed interest.

There were, however, a number of other factors which
appeared to confirm the impression within the officer
corps that, politically, they were becoming isolated from
the parent society. An important area of concern was the
changing relationship between the military and party
politics. Traditionally, this relationship had reflected
Bagehot's dictum that England was alternatively governed
by the left centre and the right centre. On the one hand,
officers had been able to base their claim to rule on
grounds which Hartington made clear:(51)

They have formed a connecting link between the advanced
party and those classes which, possessing property,
power and influence, are naturally averse to change,
and I think I may claim that it is greatly owing to
their guidance and to their action that the great and
beneficial changes which have been made in the direction of popular reform in this country, have been made
not by the shock of revolutionary agitation, but by
the calm and peaceful progress of constitutional acts.

Conversely, their self-justification had been based on
their claim to an innate right to govern. Their political
attitudes were then seen to be an extension of their central position in a society divided into the centre and the
periphery. But these two claims by the military to be
recognized as a section of the 'inevitable party men' on
the basis of their importance as heriditary legislators,
were outmoded in a changing political scene. 'Tory
democracy', of the kind advocated by Lord Randolph
Churchill, and Radicalism of the type which led Labouchere
to attack a system in which 'the opinions of a few dozen

respectable noblemen and gentlemen are likely to be
blindly accepted by the county as though they were divine
emanations',(52) took little account of the claims of
military squires. Moreover, with the development of
sharply contrasting political policies, as the changed
character of the Liberal Party emerged after the secession
of the Liberal Unionists and the passing of the Whigs, the
military appeared to be increasingly identified with only
one of the two major political parties. This was particularly marked in 1898.

This more marked ideological commitment suggested that
the military in this period, particularly in view of the
larger number of military peers than hitherto in the House
of Lords, would from time to time find itself at variance
with the 'new men' in the Commons. This increasingly
important section of society, representing the interest
of a growing middle class, were less willing than their
predecessors in the House to accept without criticism the
claim of the military to their privileges in society.
These were the politicians who had little sympathy with
the army or little understanding of its ethos. They came
from a section of society which rarely joined the army,
and, lacking in military experience, their attitude
towards the Victorian army appeared to its officers to be
founded on an antagonism which boded ill for the future.

This feeling of increasing isolation was exacerbated
by the debate in and outside of Parliament about the
relationship between the army and the civil authorities.
In part, this was a continuation of a persistent argument
about the relative power of the Commander-in-Chief and the
Secretary of State for War, an argument which lasted until
the abolition of the former post in 1895. The debate,
however, was also concerned with more fundamental issues.
As Lord Salisbury told the Queen in 1886, the army was
now so weak that it was incapable of combating a second-
class continental power,(53) a point of view which was
reiterated in the House of Lords by Wolseley who told the
House that the army would be unable, in the event of war,
even to guarantee the safety of London.(54)

The fundamental problems at stake were reviewed by a
succession of Royal Commissions which met between 1887 and
1890.(55) Each of these, perhaps not surprisingly, reached
a different conclusion. Each advocated a different remedy
to cure the military weaknesses which had become so apparent in the Russian scare of 1885 and the breakdown of
transport and supply in the Egyptian campaigns of 1882
and 1885. The first of these inquiries, the Stephen
Commission, carefully ignoring the instructions of the
Secretary for War that it was to limit its inquiry only

to the specific problem of supply, brought into question
the general administration of the military. Its conclusions were pungent and refreshingly critical.

   To rectify perceived shortcomings, the Commission sought
to remove military administration from the vagaries of
party politics and isolate it from the deficiencies of a
political system, so that the government could determine
what was actually necessary for national defence and not
what was politically possible. But because the Commission
did not want to advocate a revolutionary change in the
governmental system, its solution was based on the premise
that since other institutions in which the area of activity
had been developed from the original exercise of the royal
prerogative were now free from political interference, a
similar policy could be adopted for the military. Referring to the political neutrality of the judiciary and the
system of audit and public account, the Commission argued
that they did not see why a similar result might not be
obtained with regard to all that was technical and special
in military administration. This, it was considered, would
not in any way interfere with the fundamental principle of
complete parliamentary checks over every part of the money
expended for military purposes.(56)

   But the real crux of the problem was not so easily
solved. Efficiency in the army could in theory be ensured
by removing questions of national security from the arena
of party politics. The latter, it appeared, retarded the
establishment of any rational ordering of military priorities, preventing the development of a definite strategic
policy. Consistent changes of the political office holder
at the War Office impeded the adoption of new weapons, a
conclusion which the commission evidenced in its criticism
of Childer's reluctance to introduce a magazine-loading
rifle, despite European evidence of the advantages of such
a weapon. But any attempt to curtail the power of the
Secretary of War, or to hand a greater amount of power to
the professional soldier rather than to the amateur politician or the generalist civil servant, inevitably involved an attack on long established parliamentary principles. The question which arose was whether national
security could be guaranteed in a situation where after
transferring control over the army from the Crown to
Parliament, the military establishment was constrained in
its actions by the operation of a party system from which
it was excluded. It was a question which the Stephen
Commission could not, and did not, seek to answer.

   In many ways the ensuing debate in the press and in
public was a sterile argument. The Report which the
Stephen Commission produced was attractive to the army

because of its condemnation of Cardwell's reorganization.
Equally, it received support since it deplored a practice
in which national defence had become a political pawn.
But there was little likelihood that any political party
would surrender voluntarily its jealously guarded right
to control the administration of, and policy-making for,
the military. The reaction of officers was thus understandable for it reflected a growing feeling within the
Victorian army that it was becoming increasingly isolated
from the parent society. The traditional relationship
between the military and political élites was seen to be
slowly changing, since they were no longer linked as
hitherto by ties of kinship, shared experiences and a
common commitment to the maintenance of a privileged way
of life. Wolseley in 1890 clearly brought out the attitude of many officers at this time when he argued that,
'Party government nowadays does not mean the leading or
the endeavour to lead public opinion so much as the following of public opinion and the giving effect to it.'(57)
But, the attitude of the politician was equally rational.
An insistence on the Sovereignty of Parliament was a
fundamental characteristic of contemporary writings on the
British Constitution, so that any move to give the military
direct access to the 'people' in matters of defence policy,
was interpreted as a direct attack on the principles put
forward with much emphasis by Dicy, Austin and Bagehot.
In addition one of the concomitant effects of 'open competition' had been the development of a belief that policy-
making was more properly the province of the generalist
with his ability to see all aspects of a question, than of
the specialist with his commitment to a narrowly defined
interpretation of a problem area. The acceptance of this
conclusion, again made the attitude of the politician and
the civil servant, both of whom were generalists through
and through, a perfectly rational and understandable
reaction. Moreover, their conclusion appeared to be validated by yet another report, that of the Committee on the
Manufacturing Departments of the Army (1887), which, in
contrast with the Stephen Commission, emphasized the need
to increase the number of civilian administrators in the
government ordnance factories and to extend the scope of
their authority.(58)
    The problem which was now under discussion was a com-
plex one which went beyond the relatively simple issues
of politics. A sense of isolation was not limited to the
political sphere. Increasingly, the bureaucracy, it
appeared, was taking a larger part in the making of decisions which affected the army. The question was not simply
whether the political or military head of the War Office

was to be superordinate, but whether civil servants or
soldiers were to exercise control over military administration. The question was characteristic of the changes
which were taking place within the administrative process
of government as an increasingly professional civil service sought to enlarge the power of the bureaucracy. In
the struggle for control over this process, two of the
three competing factions, that is, the politicians and the
civil servants, seemed to many officers to be allied
against the third group of the military. There were, it
appeared, marked differences in the attitudes of these
groups towards the role of the army within society. For
the civil side, their attitude appeared to be a combination of rationalism, liberalism and sheer sentimentalism,
which, in combination, emphasized the feeling that England
had arrived at a permanent condition of security and prosperity.(59) In this situation the role of the military
was of necessity to be limited, and a tendency among the
officer corps towards an apparent militarism had to be
kept in check through the maintenance of civilian control
over the army. It was, it seemed, only less humane European powers who believed that war was an inevitable
episode of violence in a perpetual struggle for survival.
England, it was continually stressed, was not a war-like
nation but a country whose greatness and power was derived
from peaceful economic and constitutional progress.

It was against this background that the question of
political control over the military was again considered
by the second Royal Commission set up during this period
from 1885-90. The Royal Commission on Civil Establishments under the chairmanship of Sir Matthew White Ridley
rejected the general criticism of civilian administration
implicit in the Report of the earlier Stephen Commission.
In particular, it drew attention to the arguments which
were put forward to suggest that foreign, and especially
German, practice encouraged less extravagance and greater
military efficiency, before concluding that the power to
make decisions must be left to the politicians. 'It would
be contrary to the spirit and principles of our Constitution', the Ridley Commission continued 'that the Secretary
of State for War and the first Lord of the Admiralty
should be other than high Parliamentary officers, holding
Cabinet office, and owing their position to political and
not merely to professional qualifications.'(60) This conclusion was a direct rejection of the earlier recommendation that the military should be independent of politics,
and the Ridley Commission followed the argument of the
politician in urging that the role of the military was
simply that of giving professional advice to a political

head of the War Office who retained ultimate responsibility and control.

Military dissatisfaction with politicians became increasingly evident in these closing years of the Victorian army. It was not simply the effects of uncertainty about the role of the Commander-in-Chief though this often seemed to be the main issue. While the Duke of Cambridge was a convenient scapegoat who could be criticized as a hopeless reactionary who not only opposed change but who also openly quarrelled with those who proposed it,(61) the questions in issue went much deeper than this. The fundamental problem was the difficulty of establishing the boundaries of the relationship between the army and the parent society in an atmosphere in which party politics were increasingly important. What, indeed, was to be the place of the army in a democratic constitution? Hitherto, the question had not assumed the shape which it was now taking. As one of the élite institutions of society, the military had previously enjoyed a generally accepted status and a public acceptance of its claim to differential privileges. Irrespective of the actual degree of its professionalism, it had enjoyed the confidence of politicians who had deferred to its claim to expertise. Officers had been able to participate in the political process. Now the army considered itself to be under perpetual attack from politicians who not only deplored the cost of national defence but who, having excluded the military from politics, also resented the developing professionalism of the officer corps.

When officers had been 'amateurs' with no deep sense of career commitment, they had readily been identified as part of a system in which all activities within society had been controlled by a ruling class who were gentlemen first and last. But now professional soldiers, of whom Wolseley, Roberts and Kitchener were the prime examples, could no longer be so readily 'slotted' into a social system. They were, the military inferred, an embarrassment to politicians whose appeal to the public was very largely based on their apparent insistence on peaceful and constitutional progress. What other interpretation could the army place on statements such as those made in the House of Lords by the Prime Minister, Lord Salisbury, in June 1888? Defending his government against critics who had reacted to Franco-Italian strain in the Mediterranean which had resulted in the dispatch of a British fleet to Genoa to discourage potential French aggression, Salisbury made the politician's point of view very clear. Members of Parliament, he argued, should put more confidence in their front bench and less in experts. Certainly Parliament could not

surrender completely to the advice of experts. And any
accusation that the government had sought successful
budgets to the exclusion of all other considerations
including those of defence, could be dismissed, because
such charges were continually repeated against all governments in office.(62)

The conclusion drawn by many members of the military
that increasingly they were being excluded from participation in major policy-making and in decision-taking, also
seemed to be confirmed by the investigations of the third
Royal Commission established to review the relationship
between the military and the government. The Hartington
Commission set up in 1888 to investigate the army and
navy departments and their relations with the Treasury,
did not meet the demands of the military and of other reformers for a more fundamental inquiry which would decide
what the army had to do, what were the resources needed
to meet these goals and what part the army could play in
decision-making. Instead, the Commission reviewed yet
again the structural relationship of the military to the
government, and, once more, produced a different solution.
This time, the Commission concluded that the real weakness
was the office of the Commander-in-Chief. He, it was
alleged, stood between the Secretary for War and the subordinate heads of the military departments and thus prevented the former from receiving adequate professional
advice. The basic problem it appeared was not the relationship of the political head of the army to its
Commander-in-Chief, but the very existence of the military
office.(63)

The recommendations of the Hartington Commission were
ultimately of considerable importance, but initially they
did little to reassure the army and free it from its sense
of political and administrative isolation. Indeed, a
further deterioration in the relationship between the
military and the political élite can, in retrospect, be
seen to have been inevitable in the last decade of the
Victorian period when, in the autumn of 1892 following the
defeat of Salisbury's government, Campbell-Bannerman became Secretary of State for War in Gladstone's fourth and
last ministry. He inherited, despite the opposition of
the Crown, a War Office Council established on 12th May
1890 with the Commander-in-Chief as first military member.(64) But the Council rarely met, for Campbell-
Bannerman's political and administrative policy was
firmly based on his belief in liberalism, the supremacy
of the generalist and the direct and sole responsibility
of the parliamentary Minister. Consistently, his belief
in liberalism and a laissez-faire approach to problems of

national security was expressed in the House of Commons.
Not only did he oppose the creation of a General Staff
but, equally, he refused to endorse the validity of that
part of the Stanhope Memorandum which had sought to create
two Army Corps for possible use in Europe.   In his view
the superiority of foreign policy over military policy
ensured that Britain would abstain from interfering in
continental Europe.   Since the creation of Army Corps
suggested a wish to meddle in the affairs of Europe, the
scheme, he argued should be abandoned,(65) for it was not
suited to any purpose which accorded with British foreign
policy.(66)   In view of this attitude it was not perhaps
surprising that by 1894 Sir Charles Dilke could argue that
the army was in a pitiful state and that no attempt had
been made to consider the security of the Empire.(67)

Campbell-Bannerman's denunciation of the professional
expert also exacerbated the strained relations between the
military and political élites.   His endorsement of Gladstone's view that the qualities of a good administrator
were not those of a good soldier, led him to conclude that
the politician was perfectly capable of making military
decisions on the basis of the breadth of his experience.
Experts created confusion and indecision when they dis-
agreed, while in other circumstances few ministers, he
concluded, were able or willing to challenge the opinions
of their advisers.(68)   Equally, government by committee
appeared to Campbell-Bannerman to sap the foundations of
true responsibility,(69) so that his policy from 1892-5
was designed to ensure the ultimate supremacy of the
Secretary of State for War not only over the Commander-in-
Chief but also over any War Office Council or projected
defence committee.   To achieve this, the politicians and
the civil servants at the War Office carefully planned
for a further reduction in the power of the military head
of the army, plans which did not provide for consultation
of the military élite.(70)   Though, out of deference to
the known wishes of Queen Victoria the office was not
completely abolished, it was planned to vest much of the
Commander-in-Chief's power in an Army Board composed of
the principal military authorities who would in addition
have direct access to the Secretary of State for War.   The
resignation of the Duke of Cambridge was considered to be
essential to the success of the new scheme, and a considerable amount of careful political manoeuvring occurred
before the impending resignation of the Duke was announced
to the House of Commons.(71)

Although the Liberal government fell on the same day,
the succeeding administration of Lord Salisbury carried
through a programme of military reform which ensured that

the power of the Commander-in-Chief was very largely
curtailed. Broderick, the new Under-Secretary of State
for War, summarized the main principle motivating this
programme as the need to ensure 'the separate responsibility of the military heads of departments to the
Secretary of State . . . and the focussing of military
opinion by means of the Army Board'.(72) The confirming
Order-in-Council of 21st November 1895 thus gave to the
Commander-in-Chief the 'general supervision' of the military departments and tasked him with the preparation and
maintenance of schemes of offensive and defensive operations. His department substituted for a General Staff,
but the heads of military departments were to be held
directly responsible to the Secretary of State for their
particular area of concern. In addition, the functions of
the War Office Council and the Army Board were carefully
prescribed in some detail.(73)

These re-organizations of 1895 terminated the century-
old controversy about the relative status and power of the
political and military heads of the army, by removing at
one stroke of the pen the traditional powers of the
Commander-in-Chief. Henceforward, a rational-legal interpretation of power ensured that the traditional and charismatic dominance of the head of the military establishment lost most of its significance. The 1895 reforms
were the first organizational steps taken to transform
the Traditional Military into the New Military of the
twentieth century, a move which was subsequently developed
to a much greater extent by the Esher Committee and the
later Haldane Reforms.(74) Yet the 1895 re-organization
did not, and could not, completely alter overnight the
pattern of developed civil-military relationships. Attitudes were too firmly embedded. Too many senior officers
had known the old system, tending to look back with
affectionate nostalgia to days when the army seemed to
them to have enjoyed a particularly privileged position
within society. Nor were these only the 'conservative'
officers, for many of the 'radicals' who had constantly
urged reform, found that the post-1895 system only served
to increase the power of the civilian politician at the
expense of the officer corps. It was in many ways ironic
that Wolseley, who had been in the forefront of the demand
for army reforms, suddenly found on succeeding the Duke of
Cambridge, that he had inherited an almost untenable
position.

In essence, the relationship between the army and the
civil authorities in the closing years of the Victorian
period was characterized by two developments. On the one
hand, officers felt that they were excluded from

participation in decision-making at all levels of politics, as electoral and administrative reform deprived them of their previously unchallenged claim to rule as part of an ascriptive élite group. At the same time, the political attitudes of the military were increasingly affected by its developing professionalism. A growing strain between the officer corps and the politicians could not be attributed solely to changes in the pattern of electoral representation, as important as these were. Clearly there was a difference between the attitudes of the officer corps and many politicians to progress and social change. Members of Parliament who came from 'hitherto submerged groups and whose careers were bound up with individual exertion or general social and economic changes were, as a rule, emotionally and intellectually less involved in the maintenance of the social and political status quo'. (75) In contrast, the military often appeared to be the last bastion of neo-feudal attitudes and privileges. The officer corps frequently sought to preserve an outmoded, outdated attitude derived from a belief in the superiority of group membership. The virtues of bravery, discipline, obedience and patriotism were seen to be the sole prerogative of the army.

The problem, however, became more acute as officers began to follow the example of other occupational groups in objecting to lay evaluation of the expertise which underlay its claim to the possession of a monopoly in an area of occupational activity. In the closing years of the Victorian army, this reaction was an early example of military syndicalism, as officers began to stress their claim to be the sole judge on such defence issues as the size, organization, recruitment and equipment of forces. Increasingly this claim increased the danger of a collision with the civil government, particularly when the latter equally stressed that it alone had the sole right, on the grounds that defence policy was the servant of foreign policy, to decide on these issues. The two points of view were irreconcilable. Wolseley summed up the military attitude very clearly in November 1900: 'It is idle to hope that any civilian, however eminent, can have the professional training and experience required to enable him to control an army.'(76) Salisbury put forward the politician's point of view with equal clarity, when he argued that the opinion of the politician had, in a constitutional democracy, to prevail over the opinion of the military expert.(77) The fundamental differences were essentially political, not organizational in origin, and projected structural amendments, debated in an atmosphere of mistrust and recrimination did little to improve already strained civil-military relationships.

These two points of view continued to bedevil relations between the army and the parent society long after
the Victorian military had disappeared. Each was backed
by a long historical tradition. The traditional dominance
of the crown over the army had enabled the Commander-in-
Chief as the representative of the sovereign to rule the
military in the early Victorian period without challenge.
By the end of the century the doctrine of Parliamentary
control, first realised in the days of the Commonwealth,
had encouraged the evolvement of a rational legal bureaucracy in which the army, in common with other institutions in society, was subordinate to the democratic process. But the difficulty of accommodating an autocratic
institution such as the military in a parliamentary
system was not so easily overcome. Previously, accommo-
dation had been ensured through the open access to the
political system given to the army. Now this had disappeared. With what was it to be replaced?

For some officers, the answer was to be found in a
policy similar to that advocated by Lieutenant-Colonel
Henry Wilson who was inspired by 'a vision of power — a
vision of the soldier supreme, untramelled by politicians,
master of his craft and master of the army he was trying
to create'.(78) This was an extreme point of view. It
was a military attitude which resembled very closely a
point of view subsequently put forward by Von Seeckt in
the days of the Weimar Republic. 'The role of the
Reichswehr is to maintain the unity of the Reich and
those who compromise this are its enemies from whichsoever
side they come.'(79) This was an attitude of mind derived
from a developing military professionalism which led
officers to seek control over the political and administrative decisions taken within the parent society. It
was characterised by a growing mistrust of politicians.
It was symbolized by the derogatory terms used to describe politicians who became, in the words of the military,
'frocks' or 'pekins'. Sir Arthur Paget, when Commander-
in-Chief in Ireland at the time of the Curragh incident,
expressed this attitude very clearly when he asked his
recalcitrant cavalry officers if they thought he would
obey the orders of 'those dirty swines of politicians'.
(80) Growing indications of military disenchantment with
politicians were evident in the disputes which arose
between Butler and Milner in South Africa. 'Let my chief
at the War Office', argued Butler, 'tell me what to do,
and I will do it, but I will not be dragged by syndicates
in South Africa and I will not obey them, they are not my
masters.'(81) Equally, Kitchener, when Consul General in

Egypt, justified his system of personal government because 'it was dangerous to leave the really important interest of the country in the hands of inexperienced persons, swayed by outsiders' interest and moved by political wire-pullers'.(82)

The emphasis placed by these members of the officer corps on the need for the military to protect the 'national interest' created areas of considerable strain in civil-military relations. Fortunately for the main-tenance of these relations, little direct action was taken by the officer corps. Many officers were prepared to agree with Colonel Gordon-Ives (1837-1907), known to the army as the youngest officer in the Crimea, in his com-ments that his recreations in his old age were 'trying to put right the wrong and the corrupt in the country'. (83) But this did not involve the majority of officers in any direct conflict with the civil power. As Major Philip Howell, second in command of the 4th Hussars, made clear in 1914, 'The Grand military steeplechase and the like were the sole topics of conversation — all are sick to death of the subject of "Home Rule" and for weeks, I've never heard it mentioned in the mess or hunting field.'(84)

In peacetime, differences which arose could be glossed over, but the First World War soon brought home with dramatic effect the problems of civil-military relations. Kitchener's reaction when Secretary of State for War was quick and decisive. Rightly or wrongly he reverted to the traditional system whereby a single member of the military élite assumed direction of the army. Parliamen-tary debate and control by committees was constrained. Kitchener was never overruled nor even seriously chal-lenged on any important issue.(85) His word was final, even though his sweeping powers were not derived from the House of Commons. His authoritarian approach enabled him to ignore, as far as possible, all questions of politics, treating his colleagues with the usual mixture of military contempt and apprehension. 'His main idea at the Council table was to tell the politicians as little as possible of what was going on, and get back to his desk at the War Office as quickly as he could decently escape.'(86) Yet when parliamentary control was re-established, there was a very serious risk that the military élite would form a cabal which would overthrow the existing War Cabinet and enthrone a government which would be virtually the nominee and menial of the military party.(87)

This in short was the final legacy of the Victorian army. The original debate about the relative status of the Commander-in-Chief and the Secretary of State or the

right of officers to participate in politics had become,
by the end of the century, a much more controversial
argument about the place of the military in a democratic
society. After 1900, British policy, responding to
external European pressure, began to accord more importance to strategy and power, but during the lifetime of
the Victorian army, the changes which occurred were primarily designed not to meet the objective of military
efficiency, but to ensure the subordination of the army
and its officers to the control of a democratic Parliament. Defence policy, was consistently subordinated to a
vision of an international society ruled by moral law in
which the role of the army was of minimal importance. The
example of British constitutional and economic progress,
not the Victorian army, was the believed basis of Britain's
status as a world power.

8

∞∞∞∞∞∞∞∞∞∞∞∞∞∞∞∞∞∞∞∞∞∞∞∞∞∞∞∞∞∞∞∞∞∞∞∞∞∞∞

Postscript

∞∞∞∞∞∞∞∞∞∞∞∞∞∞∞∞∞∞∞∞∞∞∞∞∞∞∞∞∞∞∞∞∞∞∞∞∞∞∞

When Sir Redvers Buller landed at the Cape in October
1899, he took command of an army which, despite years
of fitful reform and Royal Commissions, was grossly
ill-prepared for the tasks which lay before it. Short
of men, ammunition and reserves of equipment, the army
was forced through indecision and defeat to improvise and
muddle through. The Second South African War was the
final campaign fought by the Victorian army. It was the
ultimate act which tested programmes of training, schemes
of organizational reform and policies of recruitment, and
immediately the army was found to be defective.

Regarded as an institution or society, this army was
undoubtedly a success. The uniforms were most distinctive.
Traditional ceremonies such as inspections, parades and
guards were elaborate and pleasing to the eye. Regulations were complex and effective. But as a fighting
machine, the British Army, despite all the eulogistic
comments made about it by L.S. Amery was, as he concluded,
'largely a sham'.(1)

In the ensuing search for a scapegoat, each of the
three interest groups involved in the War sought to put
the blame on someone else. The military were blamed for
their ineptitude in action and for their lack of fore-
sight. 'The general impression to be derived from the
whole circumstances must be that the special function of
the Commander-in-Chief, under the Order in Council of
1896, viz. "The preparation of schemes of offensive and
defensive operations", was not exercised on this occasion
in any systematic fashion.'(2) Politicians in turn were
blamed for their parsimony and for the subordination of
national security to the interests of party politics.(3)
The public was generally criticized for its neglect of the
army and for its reluctance to join its armed forces:(4)

If the terms offered are attractive only to men whose
intelligence is under-developed, it is impossible to
make them soldiers of the class required in modern
warfare, with the same amount of training that will be
sufficient for men whose mental calibre is higher at
the time when they enter the army.

But in seeking to apportion blame for the disasters in
South Africa, no one was prepared to admit that the real
reason for the failure of the Victorian military was that
it had become, by 1900, an anachronism.

As a fighting force, the Victorian army was in reality
only appropriate to a moment in history which had already
disappeared - that of an expanding Empire which depended
for its strength on an illusion of national prosperity
and on a belief in perpetual international security. In
an era of expansion, the continuing importance of the
army within the parent society was derived from a belief
in the inevitable growth without check of this empire.
The role of the regiments was clear. They were to function as a colonial army and ensure that the British
Empire continued to be 'an unchanging institution of
charitable purpose and assured income'.(5) But by the end
of the century, such assumptions no longer had any validity. The illusion of prosperity was under attack as
other nations began to industrialize and to challenge,
with their increasing competitive power, the supremacy
of Britain's economic position. Equally, the concept of
perpetual security, a concept which had encouraged a
vision of a Pax Britannica, was also questionable as the
balance of world power began to change. In an age of
strident nationalism and imperialism, Britain could no
longer be indifferent to all that happened in Europe.
Strategically, politically, and economically, Britain was
moving from a position of isolationist dominance towards
a situation in which she would be involved again in the
struggles of competing European nations.

But the army which landed in the Cape was incapable
of adapting to this new role. Fundamentally, it was the
same army as that which had fought in the Crimea. Certainly there had been changes in weapons and equipment.
No longer were the majority of the rank and file recruited
as hitherto from the ranks of a landless Irish and Scottish
peasantry. Years of spasmodic organizational reform had
amended the traditional relationships between the military
and political élites. But the ethos of the Victorian army
remained unchanged. A wide gulf continued to separate
the soldiers from their officers, who, despite the abolition of the Purchase System, were still recruited from a
limited area of society. Not that this necessarily

reduced the potential efficiency of the officer corps,
although Sir John Fisher, in looking at a comparable
situation in the Royal Navy concluded that Britain sought
to draw her Nelsons from too narrow a class.(6)   What
did follow from this traditional policy of excluding
potential officers who came from a vast reservoir of
urban middle-class talent, was the perpetuation of an
attitude which reflected social rather than functional
values.   The intellectual and moral standards of an in-
dustrialised society were rejected in favour of the conservative orthodoxy of a landed rural class.   Gentlemanly
qualities of character were preferred to the competitive
attitudes of a profit-conscious mercantile interest.
Probity, inertia and unbounded complacency rather than
enthusiasm, drive and ruthlessness were the characteristics of the Victorian officer corps, and inevitably
these became 'the classic attributes of an army about to
suffer a catastrophic defeat'.(7)

     This, therefore, was an army in which needed innovation
continued to be subservient to conservative sympathies.
These, in turn were derived from values that were functional to the professional performance of the army in its
guardian role.   Such sympathies were also perpetuated by
the historical traditions of the military.   This was an
army in which the harshness and savagery that had characterized the British Army of the eighteenth century had
largely disappeared.   It was different from Wellington's
forces which, as a brutal mass, had been subjected to a
discipline that was the most severe to be found in any
European army.(8)   Many Victorian officers, however, retained the attitudes of the flâneurs and dandies of
Regency England, looking upon active service as an unwarranted interruption of an established social life.
Others were descendants of that hard-core of 'fighting'
officers who came from a harsh rural environment, the
world of Sir Tatton Sykes (1772-1863), the good 'Old Tat'
of Surtees's novels.(9)   These were the officers who had
turned the scum of jails and the bewildered recruits of
a rural society into a disciplined fighting force.   But,
at the end of the century, both types of officers were
anachronisms.   Men hard of mind and hard of will were
still required, but, above all, it was the technical
expert, the skilled and trained professional who was
needed in the Victorian army.   In his absence, the army
seemed with some notable exceptions, to be officered by
men from the past who were resolutely opposed to technological and administrative innovation.   These were men
who resolutely rejected developments such as the introduction of the machine gun or who conducted a campaign as

if Dundas's drill book were still in force. In an era
when in other countries war was increasingly being fought
by the trained staff officer with his appreciation of
strategy and tactics, many of the officers in the Victorian
army continued to behave as independent commanders whose
campaigns were directed by intuition and conservatism.

This conservatism and intuition were, by the end of the
Victorian period, fundamental characteristics of the army.
They were attitudes of mind which, together with complacency and intertia, had been developed through a lengthy
socialization process in which the influence of the public
school was paramount. Officers, in common with other
members of the ruling élite, were men who during their
formative years had been subjected to a prolonged moulding
of character, personality and outlook. In this socialization process, the forces of technological and social
change were ignored. 'The richer Victorian England became, the more ashamed in a deep sense did she become of
the technological origin of those riches.'(10) Within
the secure world of the Victorian public school the undue
emphasis placed on the study of classics, on the acquisition of manly Christian virtues and on the idea of public
service, produced a ruling class endowed with a complacent
appreciation of their role in society. Since they were
the inevitable leaders, the need for competitiveness, the
struggle for power and the harshness of everyday life
tended to be discounted. A belief in the importance of
belonging to a team which was governed by an intricate
behavioural code that preferred honourable defeat to dis-
honest victory, discouraged the cult of personality.
Conservatism and conformity were preferred to individualism. Character not intellect was the yardstick of success.

As a part of this ruling stratum, the ethos of the
Victorian officer reflected the attitudes of the 'Establishment'. In one important respect, however, officers
differed from their contemporaries in other institutions
within society. Separated eventually from the main stream
of education by their membership of 'Army Classes' or by
their attendance at the military crammers, many officers
tended to compensate for this by over-emphasizing certain
attributes of army life. This took two main forms. For
some members of the officer corps, the evangelical tradition of a Bible in one hand and a sword in the other became a very important motivating factor. The army, in the
eyes of these officers, was identified with a puritan
ethic which implied that the parent society was morally
inferior and corrupt. For others, the social values of
the landed interest became enshrined and unchangeable, so
that the projected military image was that of an 'aristo-

cratic' institution which, in a changing world, was the
sole guardian of the traditional ethos. Here, a natural
tendency towards conservatism became a demand for the
preservation of specific values. An insistence on the
need for members of the officer corps to be 'officers and
gentlemen', then contributed to a rejection of external
pressure for enhanced professionalism, since this was
incompatible with the concept of the 'gentleman ideal'.
And even officers whose background was not that of the
landed interest were induced through their assimilation
and socialization into the privileged world of the regiment to accept these social values. There were always
exceptions. Some non-conforming officers achieved high
rank and a considerable reputation, as did Wilson and
Kitchener. Most, however, realizing their inability to
accept the constraints of regimental life, preferred to
leave the army, a decision which increased still further
the complacent homogeneity of those who remained in the
military establishment.

It was this complacency which ultimately epitomized
the ethos of the Victorian army. In maintaining this
attitude, officers were not alone, for much of their value-
system was derived from the common educational experience
and life-style which they shared with other members of the
ruling class. It was a complacency which also symbolized
the attitude of many 'new' politicians who came from
hitherto submerged groups but who were prepared to accept
a national belief in Victorian prosperity and security.
And since this was a society which was prepared to ignore
the dissenting attitudes of the 'lower orders', complacency seemed to assume the mantle of a national characteristic. From time to time, it was shattered. The Crimea,
the Mutiny and Majuba were rude awakenings, but a long
succession of colonial victories against ill-armed and
badly-organized native forces soon eroded these memories.

Yet while the complacency of the politician and the
public could be accepted, even if it were to be deplored,
the failure of the Victorian army to respond to changes
in world politics was less excusable. Complacency in this
area of society was a luxury which could not be afforded.
Technological changes which were both a cause and effect
of the expansionist ambitions of European nation states,
were already producing changes in the balance of world
power. Increasingly, the geographical pivot of history
was shifting away from the rimland back toward the traditional continental area of conflict.(11) Industrializa-
tion, which had encouraged a growing mobility in terms of
sea power, was now producing a comparable development
in the mobility of land power. There were strategists

who in ignoring the significance of this change, maintained that the role of Britain in the world of power-
blocs was still that of a maritime power. As Spenser
Wilkinson argued during a discussion of Mackinder's views
on the geographical pivot of history,(12)

> You have had in the west of Europe a small island which,
> having attained to its own political unity, and having
> in the conflict for its own independence developed its
> sea-power, has been able to affect the marginal
> regions, and to acquire the enormous influence which
> was revealed to us — the British Empire. . . . My own
> belief is that an island state like our own can, if it
> maintains its naval power, hold the balance between
> the divided forces which work on the continental area,
> and I believe that this has been the historical func-
> tion of Great Britain since Great Britain was a United
> Kingdom.

But sea power was no longer the sole factor of importance in a European situation where the possession of
industrial power and mobility by land gave immeasurable
advantages to the continental nations. Already, the
inability of Palmerston to influence the outcome of the
1864 dispute between Prussia and Denmark had shown up the
limitations of gunboat diplomacy. A powerful professional
army rather than a small colonial force protected by a
large navy was now needed as the instrument of British
policy in this changing scene.

But for the reasons which have been stated, the
Victorian military was not designed to fill this role.
It was an army which had been created for the primary
purpose of meeting the needs of Britain as an outward-
looking oceanic nation. Its professionalism was suspect.
The national preference for the gifted amateur rather
than the committed professional in all areas of the public
service distorted military training programmes. It
encouraged an acceptance of charismatic leadership as a
basis of authority and for promotion. A pragmatic attitude towards regulations and records was paralleled by a
reluctance to consider the effect of changes in weaponry
and equipment. The undue emphasis which was placed on a
'traditional' approach to ethical and social values, encouraged an acceptance of unsophisticated standards of
courage, loyalty and ability. Equally, the organization
of the military implied a rejection of the bureaucratic
model which was an essential characteristic of the continental mass army. The planned approach of European armies
towards military goals and decision-making, was replaced
among Victorian officers by a tendency to consider the
maintenance of established norms and values as having more

importance than ends and means. Even administrative
reorganization did little to change the self-image of the
officer corps with its rigid sense of 'form' and its
predilection for a complex social etiquette and an unchanging way of life. In its political relationship to
the civil power, the military similarly sought to maintain
its privileged position as an inevitable part of a permanent ruling class. Despite constant changes in the composition of the political élite, the officer corps endea-
voured to preserve its special relationship with the civil
power. Thus an innate military conservatism re-emphasized
the importance of the army to society as the guardian of
stable values within a changing political climate, an
attitude which bred in the Victorian officer a feeling
of responsibility for the well-being of the state.

   In all these characteristics, the military was the
mirror of the parent society. The army was not created
in isolation. It did not function as an independent
organization. Its professionalism, its administration and
its political attitudes reflected the ethos of a part, if
not the whole, of the society within which it operated.
If, at the end, the army was not prepared for the Second
South African War, then this echoed a more general and
national unwillingness to face up to the fact that in
troublesome days, isolation, though splendid, was not a
desirable situation.(13) Britain by 1900 could no longer
continue to bask in the reflected glory of the largest
empire in the world protected by the most powerful navy.
Within two years, the alliance with Japan showed that in
a world which was quickly contracting under the spur of
technological development, the traditional attitude of
the Victorian toward armed forces was outmoded. A small
volunteer army, primarily designed to implement a guardian
role in Britain and the Empire, was inadequate in a
European situation.

   To a considerable degree, this inadequacy was exposed
during the campaigns in the Cape. Consistently, the
successes of the Boers suggested that the Victorian army
could not adapt to change. A military which for sixty-
four years had spread the sphere of British influence
throughout the world, was now seemingly unable to cope
with its tasks in a changing society. This anachronistic
armed force could be bitterly criticized for its shortcomings. Yet these shortcomings and this inadequacy were
a more general reflection of contemporary society. In
the absence of a military caste, the Victorian army was
the mirror of the parent society and, if the army was
less than successful, then the final conclusion must be
that this society was primarily to blame. In short, the

relationship between the army and Victorian society was
one of total interdependence in which military successes —
and failures — were ultimately the responsibility of the
civil government and of the public in general. When the
army had achieved victory in a succession of campaigns
fought against ill-armed and badly-organized native opponents, the population had basked in the reflected glory.
War had been a distant adventure. The Cape, however,
brought home with dramatic effect the extent to which an
army that had been the pride of Victorian society, was
now ineffective. Once again, Britain had the army it
deserved.

Notes

CHAPTER 1   THE IMPACT OF DEFEAT

1   J.W. Mackail and Guy Wyndham, 'Life and Letters of
    George Wyndham', London, 1926, vol. i, p. 361.
    Wyndham to his mother, 6 October 1899.
2   A Boer War: The Military Aspect, 'Blackwood's
    Magazine', August 1899, p. 265.
3   'Minutes of Evidence taken before the Royal Commission
    on the War in South Africa', vol. ii (Cd. 1791, 1903)
    XLI, 1904, Evidence of Lord Lansdowne at 21237.
4   Sir Phillip Magnus, 'Kitchener, Portrait of an
    Imperialist, London, 1958, pp. 171-2, citing a letter
    from Kitchener.
5   The Military Situation in the Transvaal, 'Spectator',
    10 August 1899, pp. 240-1.
6   'Report of His Majesty's Commissioners appointed to
    inquire into the Military Preparations, and other
    matters connected with the War in South Africa' (Cd.
    1789, 1903) XL, 1904. 'Minutes of Evidence taken
    before the Royal Commission on the War in South
    Africa', vol. i (Cd. 1790, 1903) XL, 1904 and vol. ii
    (Cd. 1791, 1903) XLI, 1904. 'Appendices to the
    Minutes of Evidence taken before the Royal Commission
    on the War in South Africa' (Cd. 1792, 1903) XLIII,
    1904.
7   J.F.C. Harrison, 'The Early Victorians 1832-51',
    London, 1971, p. 104.
8   See 'Hansard', vol. 350, p. 1143. Also see, 'Report
    of committee of general and other officers of the army
    on re-organization' (Cd. 2791, 1881) XXI, p. 30.
9   'Hansard', vol. 129, 5 February 1904, p. 552. Speech
    by Captain Cecil Norton.

10 Sir John W. Fortescue, The Vicissitudes of Organized Power, (the Romanes Lecture, 22 May 1929), in 'The Last Post', Edinburgh, 1934, p. 49.
11 Theodore von Sosnosky, 'England's Danger: the Future of British Army Reform', London, 1901, p. 80.
12 For a further analysis of the characteristics of militarism, militocracy and militolatry, see Stanislav Andreski, 'Military Organization and Society', London, 1968, pp. 184 ff, and Andrew Vagts, 'A History of Militarism', New York, 1937.
13 See Sir Charles Trevelyan, 'The British Army in 1868', London, 1868, and E.B. De Fonblanque, 'Treatise on the Administration and Organization of the British Army', London, 1858.
14 Among the most influential examples of this campaign literature were George Younghusband's 'The Relief of Chitral', London, 1895, and G.W. Steevens' 'With Kitchener to Khartum', Edinburgh, 1898. The latter book by 1900 was in its twenty-first edition.
15 Gaetano Mosca, 'Elementi di scienza politica', translated by H.D. Kaln and edited by A. Livingston as 'The Ruling Class', New York, 1939, p. 233.
16 See Charles C. Moskos Jnr, Armed Forces and American Society: Convergence or Divergence in Moskos (ed.), 'Public Opinion and the Military Establishment', Beverley Hills, 1971.
17 See 'Murray Papers', W.O. 3/454, pp. 36-221 and 3/455, pp. 155-60 for further details of the Court Martial of the Earl of Cardigan, and 'Murray Papers', W.O. 80/13, for 'Observations on the Practice of Duelling and the Introduction of Courts of Honour in the Army with a View to Restraining it', 1843.

CHAPTER 2   OFFICER RECRUITMENT

1 M. Janowitz, 'The Professional Soldier',New York, 1960, p. 80.
2 See The Northcote-Trevelyan Report on the Home Civil Service, 'Parliamentary Papers 1854-5', XXVII, 1713.
3 W.L. Burn, 'The Age of Equipoise', London, 1964, p. 261.
4 See A.W. Kinglake, 'The Invasion of the Crimea', London, 1877, vol. iii, pp. 169-70.
5 G.R. St Aubyn, 'The Royal George', London, 1963, p. 200 and p. 86.
6 Sir John W. Fortescue, Hugh, first Viscount Gough, in 'The Last Post', pp. 167-73.
7 'The Times', 24th October 1840.

8   Lord Stanmore, 'Sidney Herbert', London, 1906, vol. ii,
    p. 101. Also see the speech by Sidney Herbert in the
    House of Commons, 4 March 1856, 'Hansard', vol. 140,
    p. 1,843.
9   'Hansard', vol. 157, 6 March 1860, p. 48.
10 M. Curling (ed.), 'Recollections of Rifleman Harris',
    London, 1848, reprinted in J. McGuffie (ed.), 'Rank
    and File', London, 1964, p. 145.
11 See Raoul Giradet, 'La société militaire dans la
    France contemporaine (1815-1939)', Paris, 1953, p. 54.
12 'Saturday Review', 8 March 1856.
13 R.H. Gronow, 'The Reminiscences and Recollections of
    Captain Gronow', London, 1892, vol. ii, pp. 268-9.
14 'The Times', 20 August 1857.
15 'The Times', 5 March, 1858.
16 'Report of the Commissioners Appointed to Inquire into
    the System of Purchase and Sale of Commissions in the
    Army, with Evidence and Appendix', 1857, Sess 2 (2267),
    XVIII, Question 4607, Evidence of Sir Charles
    Trevelyan.
17 Jenifer Hart, Sir Charles Trevelyan at the Treasury,
    'English Historical Review', LXXV, no. 294, 1960,
    pp. 15-17. Trevelyan's correspondence with Russell,
    1848-9.
18 Edward Hughes, Civil Service Reform 1853-55, 'History'
    (New Series), XXVII, 1942, p. 83.
19 For a further discussion of the political reaction
    to the demand for 'open competition', see Donald
    Southgate, 'The Passing of the Whigs, 1832-1886',
    London, 1962, pp. 204 ff.
20 W.M. Thackeray, 'The Book of Snobs', London, n.d.,
    Chapter IX, On some military snobs; Chapter X,
    Military snobs and Chapter XXIX, A visit to some
    country snobs.
21 Burn, op.cit., p. 15.
22 W.W. Rostow, 'Stages of Economic Growth', Cambridge,
    1960.
23 Benjamin Disraeli, Speech at Shrewsbury, 9 May 1843.
    Quoted in R.J. White (ed.), 'The Conservative Tradi-
    tion', London, 1950, pp. 174-5.
24 For a detailed analysis of changes which occurred in
    the landed interest during this period, see F.M.L.
    Thompson, 'English Landed Society in the Nineteenth
    Century', London, 1963, and Harold Perkin, 'The
    Origins of Modern English Society, 1780-1880, London,
    1969.
25 Thompson, op.cit., pp. 4 ff. An alternative interpre-
    tation of this structure, which stresses that the
    solidarity of rural society was an illusion, is given

by E.J. Hobsbawn and George Rudé, 'Captain Swing',
London, 1969, pp. 17 ff.
26 Thompson, op.cit., quoted at p. 6.
27 J. Bateman, 'The Great Landowners of Great Britain
and Ireland' (4th edn), London, 1883 (reprinted by
Leicester University Press, 1971). Bateman's work
was based on 'Parliamentary Papers', 1874, LXXII,
Return of Owners of Land, 1872-3 (The New Domesday
Survey).
28 Charles de Montalembert, 'The Political Future of
England', London, 1856, p. 85.
29 See P.A. Bromhead, 'The House of Lords in Contemporary
Politics 1911-1957', London, 1958, p. 25.
30 'The Letters of Queen Victoria, 1st Series', London,
1908, vol. ii, p. 480.
31 J.T. Ward, 'East Yorkshire Landed Estates in the
Nineteenth Century', East Yorkshire Local History
Society Series, no. 23, York, 1967, p. 50.
32 Thompson, op.cit., pp. 198-9.
33 Compiled from the Army List, 1838 and 'Burke's Peerage
and Baronetage 1838', London, 1838.
34 See E.C. Barber, 'The Bourgeoisie in Eighteenth Century
France', London, 1964.
35 J.B. Burke, 'The Vicissitudes of Families', 2nd series,
London, 1861, p. 6.
36 P.E. Razzell, Social Origins of Officers in the Indian
and British Home Army, 'British Journal of Sociology',
vol. XIV, September, 1963, pp. 248-60. These figures
are based on data for 1830 but there are no significant
changes in the source of recruitment between this date
and 1838.
37 Ibid., p. 254.
38 Ibid.
39 J. Caird, 'The Landed Interest and the Supply of Food',
London, 1878.
40 De Fonblanque, 'Treatise on the Administration and
Organization of the British Army', pp. 236-7.
41 See Edward McCourt, 'Remember Butler', London, 1967,
passim.
42 Ibid., p. 12.
43 'The Times', 10 August 1849.
44 E.G. Bulwer Lytton, 'England and the English', London,
1887, pp. 29-31.
45 See W.H. Aydellotte, Patterns of National Development
in R. Appleman et al. (eds), '1859, Entering an Age
of Crisis', Bloomington, Ind., 1959, pp. 118 ff.
46 See W.L. Guttsman, 'The British Political Elite',
London, 1965, p. 168.

47 See Michael Glover, 'Wellington as a Military Comman-
    der', London, 1968, p. 209.
48 Sources: Army List, 1838 and 'Burke's Peerage and
    Baronetage', 1838.
49 C.B. Otley, Social Affiliations of the British Army
    Elite, in Jacques van Doorn (ed.), 'Armed Forces and
    Society', The Hague, 1965, p. 89.
50 Razzell, op.cit., p. 254.
51 W.E. Cairnes, 'Social Life in the British Army',
    London, 1900, p. 26.
52 'Report on Training Officers for the Scientific Corps',
    1856, Evidence of General Sir Hew Ross at p. 322.
53 E.G. French, 'Good-bye to Boot and Saddle', quoted
    in E.S. Turner, 'Gallant Gentlemen', London, 1956,
    p. 242.
54 Lady Wantage, 'Lord Wantage V.C., K.C.B.', London,
    1907, pp. 376-7.
55 Steevens, 'With Kitchener to Khartum', p. 113.
56 General Sir George Higginson, '71 Years of a Guards-
    man's Life', London, 1916, p. 74.
57 For the German figures, see M. Janowitz, op.cit.,
    p. 96.
58 'Who Was Who', 1897-1916, p. 94.
59 R. Cunningham, 'Conditions of Social Well-Being',
    London, 1878, p. 328.
60 See G.A. Denison 'Notes of My Life 1805-1878', Oxford,
    1878, pp. 1-3. For a description of the part played
    by the family in politics, notably by John Viscount
    Ossington, see Donald Southgate, 'The Passing of the
    Whigs 1832-1886'.
61 J.S. Mill, 'Letters' (ed. H.S.R. Elliot), London,
    1910, vol. i, p. 205.
62 Salazar, 'El Pensamiento de la revolucion nacional',
    Buenos Aires, 1938, chapter V, Elogio de las
    virtudes militares, pp. 118-22.
63 Bernard Cracroft, The Analysis of the House of
    Commons, or Indirect Representation, in 'Essays on
    Reform', London, 1867, pp. 156-65 at p. 158.

CHAPTER 3   THE PURCHASE SYSTEM

1 W. Cobbett, 'The Progress of a Ploughboy to a Seat in
   Parliament', quoted in J. McGuffie, 'Rank and File',
   p. 154.
2 Sir Charles Trevelyan, 'The Purchase System in the
   British Army', London, 1867, p. 2.
3 Ibid.

4   H. Taine, 'Notes on England (Notes sur L'Angleterre,
    1872)', translated by E. Hyams, reprinted in W.L.
    Guttsman (ed.), 'The English Ruling Class', London,
    1969, p. 38.
5   Prussian Cabinet Order of 9th September, 1808. Cited
    in K. Friedlander, 'Kriegs-Schule', Berlin, 1850,
    p. 225. As Correlli Barnett points out in The Educa-
    tion of Military Elites, 'Journal of Contemporary
    History', vol. 2, no. 3, 1967, at p. 19, it was the
    inner élite of one of the most caste-bound and pri-
    vileged officer corps in Europe — the Prussian —
    which first adjusted to the demand for 'professional'
    officers.
6   Trevelyan, op.cit., p. 2.
7   Max Weber, 'The Theory of Social and Economic Organi-
    zation', trans. A.M. Henderson and ed. Talcott
    Parsons, Chicago, 1947, p. 347.
8   'Report of the Royal Commission on Naval and Military
    Promotion', 1838, p. 285.
9   C.H. Firth, 'Cromwell's Army', London, 1962, p. 18.
10 Clarendon, 'Rebellion', vi, quoted in Firth, op.cit.,
    p. 19.
11 See Sir Robert Biddulph, 'Lord Cardwell at the War
    Office: a history of his administration, 1868-1874',
    London, 1904, p. 81.
12 H. Bryerley Thomson, 'The Choice of a Profession',
    London, 1857, p. 189.
13 Memorandum of General Officers, 1821.
14 49 Geo. III, c 126 se 8, 'Sale of Offices Act, 1809'.
    The Crown, however, expressly reserved its discretion-
    ary right to continue the sale of army commissions.
15 Calculated from a War Office Return dated April 1838.
    Cited in 'Report of the Royal Commission on Naval and
    Military Promotion', 1838, p. 104.
16 M.J. Higgins, 'Letter on Army Reforms', London, 1855,
    p. 3.
17 Calculated from 'Hart's Army List', 1856, and 'Report
    of the Purchase Commission', 1857, XVIII, pp. 409-18.
18 'Hansard', vol. 140, 4 March 1856, pp. 1,821-2.
19 'Report of the Purchase Commission', 1857, p. xxv.
20 Ibid., p. xxiv.
21 Ibid. Evidence of Major-General Sir Charles Yorke,
    Military Secretary to the Commander-in-Chief, at p. 6.
22 Ibid., Evidence of Charles Hammersley of Cox's, the
    Army Agents, at p. 49.
23 See 'Hansard', vol. 140, 4 March 1856. Speech of
    General Sir De Lacy Evans, p. 794.
24 'Report of the Purchase Commission', 1857. Evidence
    of Colonel Lord West at p. 155.

25   Ibid., p. xxi and Evidence of Lord Panmure at p. 211.
26   Cecil Woodham-Smith, 'The Reason Why', London, 1953, p. 26.
27   Christopher Hibbert, 'The Destruction of Lord Raglan', London, 1963, p. 29.
28   'Report of the Purchase Commission', 1857, Evidence of Charles Hammersley at p. 49.
29   For the regulations on the issue of half pay, see 'War Office Papers', 33/19, pp. 141-73.
30   'Report of the Purchase Commission', 1857, p. 155.
31   See S.F. Scott, The French Revolution and the Professionalization of the French Officer Corps, 1789-1793, in M. Janowitz and J. van Doorn (eds), 'On Military Ideology', Rotterdam, 1971, p. 9.
32   Turner, 'Gallant Gentlemen', p. 226.
33   Calculated from the Army List, 1838 and 'Report of the Royal Commission on Naval and Military Promotion', 1838, p. 104.
34   Calculated from 'Report of the Purchase Commission', 1857, Appendix XII, p. 389.
35   Ibid., p. 193.
36   Ibid., p. 132.
37   Ibid., Appendix VIII.
38   Ibid., Evidence of Major-General Sir C. Yorke at pp. 7 and 39.
39   Calculated from War Office Returns dated 3 July 1856, published in 'Report of the Royal Commission on Purchase', 1857, as Appendix III.
40   For the revised regulations and the discussions which preceded their introduction, see 'War Office Papers' WO 43/90.
41   See Lord Stanmore, 'Sidney Herbert', vol. ii, p. 101.
42   See Archibald Forbes, 'Colin Campbell', London, 1895, pp. 104-5.
43   Max Weber, 'The Protestant Ethic and the Spirit of Capitalism', London, 1930, p. 261.
44   Trevelyan, op.cit., p. 7.
45   De Fonblanque, 'Treatise on the Administration and Organization of the British Army', pp. 128-9.
46   Captain G. Wrottesley (ed.), 'Military Opinions of Sir John Burgoyne', London, 1859, p. 53.
47   Lord Coleridge, 'This For Remembrance', London, 1925, p. 69.
48   'The Times', 8 March 1855.
49   For a more detailed appreciation of this action, declared by the Duke of Cambridge to be 'a rout so complete and disastrous that it is almost unparallelled in the long annals of our Army', see Sir W.F. Butler, 'The Life of Sir George Pomeroy-Colley', London, 1899,

Sir Ian Hamilton, 'Listening for the Drums', London, 1944, and Brian Bond, The South African War, 1880-1, in Brian Bond (ed.), 'Victorian Military Campaigns', London, 1967.

50 The position in the mid-Victorian Civil Service is discussed in Edward Romilly, 'Promotion in the Civil Service', London, 1848.

51 'Queen's Regulations for the Army' (1850 edn), Chapter 267.

52 Ibid.

53 Ibid., Chapter 189, para. d.

54 Trevelyan, 'The British Army in 1868', p. 24.

55 Quoted in Turner, 'Gallant Gentlemen', p. 164.

56 Trevelyan, 'The British Army in 1868', p. 24 and Trevelyan, 'The Purchase System in the British Army', pp. 1-2 and p. 3.

57 Memorandum of the Duke of Wellington, 1830. Cited in S.H. Stocqueler (pseudonym for Joachim Heywood Siddons), 'A Personal History of the Horse-Guards', London, 1873, p. 153. The views of other senior officers in the 1830s were similar to those of Wellington. See 'Report of the Royal Commission on Naval and Military Appointments', 1838, Evidence of Generals Gordon, Blakeney and Lord Fitzroy Somerset.

58 'Hansard', vol. 140, 4 March 1856, p. 1,791-850.

59 Letter from the Duke of Wellington to Lord Hill. Cited in 'Report from the Select Committee on the Establishments of the Garrisons and on the Pay and Emoluments of Army and Naval Officers, with Minutes of Evidence and Appendix', 1833, House of Commons, 12 August 1838, p. 274.

60 Quoted in Captain Owen Wheeler, 'The War Office Past and Present', London, 1914, p. 100

61 De Fonblanque, op.cit., pp. 258-9.

62 Baron F.P.C. Dupin, 'Military Forces of Great Britain', London, 1857, p. 23.

63 'Hansard', vol. 109, p. 650.

64 Calculated from evidence given by De Fonblanque, op.cit., p. 260.

65 'Report of the Select Committee on Naval and Military Appointments', 1833, Evidence of the Duke of Wellington at p. 274.

66 'Report of the Royal Commission on Naval and Military Promotion', 1838, p. 318.

67 Trevelyan, 'The Purchase System in the British Army', pp. 42-3.

68 Thackeray, 'The Book of Snobs', p. 102.

69 Adam Smith, 'The Wealth of Nations', London, 1933, book I, p. 161.

70   See 'Object of Mess Allowances', 1840, W.O. 3/433,
     p. 390 and 'Committee Report on Mess Allowances',
     1882, W.O. 33/40,pp. 5-12.
71   Max Weber, 'The Theory of Social and Economic Organi-
     zation', p. 351.
72   Trevelyan, 'The British Army in 1868', p. 7.
73   Cunningham, 'Conditions of Social Well-Being', p. 330.
74   Trevelyan, 'The Purchase System in the British Army',
     pp. 7-8.
75   Speech by Lieutenant-Colonel Anson in the House of
     Commons on the Army Bill of 1870. Cited in Biddulph,
     'Lord Cardwell at the War Office', p. 231.
76   Queen Victoria to Childers, November 1880. Spencer
     Childers, 'The Life and Correspondence of the Rt Hon.
     Hugh C.E. Childers, 1827-1896', London, 1901, vol. ii,
     p. 73. (Hereafter referred to as 'The Life of
     Childers'.)
77   Biddulph, 'Lord Cardwell at the War Office', p. 116.
78   J.W. Fortescue, 'A History of the British Army',
     London, 1930, vol. xiii, p. 560.
79   For the debates on the defeated Army Regulation Bill
     of 1871 which examined these potential changes see,
     'Hansard', vols 204, 207.
80   Anon., The Army and Democracy, 'Fortnightly Review',
     March 1886, p. 340.
81   Field Marshal Sir William Robertson, 'From Private
     to Field Marshal', London, 1921, pp. 29-31.
82   W.E. Cairnes, 'Social Life in the British Army', p. 14.
83   Ibid., p. 7.
84   John Baynes, 'Morale', London, 1967, p. 29.
85   Cairnes, 'Social Life in the British Army', pp. xviii
     to xix.
86   'Report of the Committee to inquire into the Nature
     of the Expenses Incurred by Officers of the Army',
     Cd. 1421, 1903, X, pp. 7-8.
87   Baynes, op.cit., p. 29.
88   Peter Laslett, 'The World We Have Lost', London, 1971,
     pp. 227 ff.
89   Cairnes, 'Social Life in the British Army', p. 36.
90   'Hansard', vol. 38, 19 June 1896, pp. 1481-4.
91   'Report of the Committee to Inquire into the Nature
     of the Expenses Incurred by Officers of the Army',
     1903, X, pp. 7-8.
92   'Report of the Purchase Commission', 1857, p. xxiv.
     The position did not change after 1871. See W.S.
     Hamer, 'The British Army, Civil-Military Relations
     1885-1905', London, 1970, p. 17.
93   Cairnes, 'Social Life in the British Army', pp. xiii-
     xiv.

94 G.C. Brodrick, A Nation of Amateurs, 'Nineteenth
    Century', October 1900.
95 Fisher to Sir Francis Knollys, February 1904. A.J.
    Marder (ed.), 'Fear God and Dread Nought: The
    Correspondence of Admiral of the Fleet Lord Fisher
    of Kilverstone', Cambridge, Mass., 1952, vol. i,
    pp. 300-1.
96 R.A.L. Pennington, Army Reform from a Battalion
    Point of View, 'Fortnightly Review', February 1901,
    p. 326.
97 Lieutenant-Colonel Colin Mackenzie, 'Storms and Sun-
    shine of a Soldier's Life', London, 1884.
98 G.W. Steevens, 'With Kitchener to Khartum', p. 75.
99 C. Ballard, 'Smith-Dorrien', London, 1931, pp. 48-9.
100 A.J. Barker, 'Townshend of Kut. A Biography of
    Major-General Sir Charles Townshend KCB, DSO',
    London, 1967, p. 49.
101 E.L. Woodward, 'The Age of Reform', Oxford, 1938,
    p. 258.
102 'Hansard', vol. 2, 7 March 1892, p. 202.

CHAPTER 4   PROFESSIONAL EDUCATION

1 Abraham Flexner, Is Social Work a Profession?, in
   'Proceedings, National Conferences of Charities and
   Corrections', New York, 1915, pp. 576-90, and
   William J. Goode, The Theoretical Limits of Profes-
   sionalization, in Amitai Etzioni (ed.), 'The Semi-
   Professions and their Organization: Teachers, Nurses,
   Social Workers', New York, 1969, pp. 216-313.
2 For example, Howard S. Becker, The Nature of a Profession, in 'Education for the Professions', Chicago,
   1962, pp. 27-46.
3 Denison to Sir Charles Wood, Secretary of State for
   India, 21 October 1864. Quoted in J.M. Compton, Open
   Competition and the Indian Civil Service 1854-1876,
   'English Historical Review', vol. 83, 1968, p. 273.
4 'Report of H.M. Commissioners on Revenues and Management of Certain Colleges and Schools, studies pursued
   and instruction given' (3288), vol. i, 'Report', 1864,
   xx, p. 66, ('Report of the Public Schools Commission').
5 G.C.M. Birdwood, 'Competition and the Indian Civil
   Service', London, 1872, p. 10.
6 G.W. Steevens, 'With Kitchener to Khartum', p. 91.
7 M.D. Feld, The Military Self-Image in a Technological
   Environment, in M. Janowitz (ed.), 'The New Military',
   Chicago, 1967, p. 163.

8   Ernest Greenwood, Attributes of a Profession, 'Social
    Work', 2, 3, 1957, p. 45.
9   For a fuller account of this Society, see Jay Luvaas,
    'The Education of an Army', London, 1965.
10 Published in the 'Edinburgh Review', XXXV, 1821,
    pp. 377-409.
11 Lieutenant-General Sir John F. Burgoyne, 'Army
    Reform', London, 1850, pp. 10-16.
12 Major-General Sir John Mitchell, 'Thoughts on Tactics
    and Military Organisation', London, 1838, pp. 1-2.
13 Diocledes, On Presence of Mind, 'United Services
    Magazine', 1848, 6, p. 39.
14 The British Army, Past, Present and Future, 'United
    Services Magazine', 1842, 1, p. 22.
15 Michalena Vaughan and Margaret Scotford Archer, 'Social
    Conflicts and Educational Change in England and France,
    1789-1848', Cambridge, 1971, p. 51.
16 E. Thring, 'Education and School', Cambridge, 1864,
    p. 94.
17 Attorney General v Whitely (1805), 'Vesey's Chancery
    Reports', xi, p. 242.   Attorney General v Earl of
    Mansfield (1826), 'Russell's Reports in Chancery',
    II, p. 501.
18 Thomas Arnold, 'Miscellaneous Works', London, 1845,
    p. 230.
19 An advertisement for 'Queen Square Academy' claimed
    'Young Gentlemen are carefully instructed in every
    branch of useful and polite learning, and qualified,
    for the University, the Army, the Navy and the count-
    ing house.' 'The Times', June 22, 1815.
20 Ian Worthington, 'Antecedent Education and Officer
    Recruitment: the Origin and Early Development of the
    Public School-Army Relationship'. Paper to the
    British Inter-University Seminar on Armed Forces
    and Society, Hull, 1974.
21 Brian Simon, 'Studies in the History of Education,
    1780-1870',London, 1965, p. 79.
22 V. Knox, 'Remarks on the tendency of certain clauses
    in a Bill now pending in Parliament to degrade
    Grammar Schools', London, 1820, p. 1.
23 Nicholas Hans, 'New Trends in Education in the
    Eighteenth Century', London, 1951, Table III at p. 119.
24 Cardinal Newman, 'The Idea of a University', London,
    1852, Preface.
25 See 'Report From a Select Committee on the Indian
    Civil Service, (1854)', 1854-5, lv, 34, p. 3.
26 Letter of 9 May 1836, quoted in T.W. Bamford, 'Thomas
    Arnold', London, 1960, p. 120.

27 J. Fitch, 'Thomas and Matthew Arnold and their
    Influence on English Education', London, 1897,
    p. 141.
28 Royal Warrant of 30 April 1741. 'State Papers',
    44/184 and 41/36.
29 A. Adamson, 'English Education, 1789-1902', Cambridge,
    1964, p. 45.
30 See J. Ashby, 'Technology and the Academies', London,
    1959, passim.
31 M. Brewster, Decline of Science in England, 'Quarterly
    Review', 1830, xliii, p. 307.
32 Herbert Spencer, 'Education: Intellectual, Moral
    and Physical', London, 1861, p. 39.
33 'Regulations for the Admission of Gentlemen Cadets
    into the Royal Military Academy', 1838.
34 'Report on Training Officers for the Scientific Corps',
    1856, p. 1i.
35 For the revised regulations of 1835, see W.O. 44/451.
36 'Report on Training Officers for the Scientific Corps',
    1856. Evidence of Professor Sylvester, Professor of
    Mathematics, at p. 414.
37 Ibid. Evidence of Major-General W.D. Jones at p. 394.
38 Ibid. Evidence of Colonel Sandham, p. 380, Professor
    J.J. Sylvester, p. 414, and Lieutenant-Colonel M'Kerlie,
    p. 358.
39 Ibid., pp. 314-15.
40 Ibid. Evidence of Major Charteris, p. 325, of General
    Sir Hew Ross, p. 321, and of General Sir Howard
    Douglas, p. 315.
41 Karl Demeter, 'The German Officer Corps in Society
    and State 1650-1945', London, 1965, p. 279.
42 'Report of the Royal Commission of Naval and Military
    Promotion', 1838, Evidence of Lieutenant-Colonel
    Edward Mitchell at p. 41.
43 See R.H. Thoumine, 'Scientific Soldier, A Life of
    General Le Marchant 1766-1812', Oxford, 1968, pp. 80-
    98.
44 Brigadier Sir John Smyth, 'Sandhurst: The History of
    the Royal Military Academy Woolwich, The Royal Military
    College, Sandhurst, and the Royal Military Academy
    Sandhurst 1741-1961', London, 1961, p. 58.
45 E.C. Masland and M. Radway, 'Soldiers and Scholars',
    Princeton, N.J., 1957, pp. 77-8.
46 Thoumine, op.cit., p. 64.
47 Quoted in Brevet-Major (later Lieutenant-General Sir
    Alfred) Godwin-Austen, 'The Staff and the Staff
    College', London, 1927, p. 102.
48 'Report of the Royal Commission on Naval and Military
    Promotion', 1838, p. 104.

49   'Hansard', third series, vol. 103 (2), p. 963.
50   'Quarterly Review', CLXVI, September 1848, p. 422.
51   'Parliamentary Papers', 1849, XXXII, p. 532.
52   Worthington, op.cit., p. 9. Also see W.O. 33/3A/73-76, and W.O. 33/4B/24 and 26.
53   Calculated from War Office Return of 30 June 1838.
54   'Report on Training Officers for the Scientific Corps', 1856, p. xxvii.
55   Ibid.
56   Undated letter from Colonel Charles Crawford to Colonel John Le Marchant, 1798. Cited in D. Le Marchant, 'The Memoirs of the late Major-General Le Marchant', London, 1841, p. 83.
57   Quoted in Godwin-Austen, op.cit., p. 155.
58   Lord Raglan to the Duke of Newcastle 15 January 1855 ('Raglan Crimean Papers' MM184 R to N) quoted in Hibbert, 'The Destruction of Lord Raglan'. Raglan's letter is in reply to his critics who had complained of the general inefficiency of the Staff in the Crimea, and of the 'aristocratic hauteur, incivility and God knows what besides' of Raglan's personal staff.
59   See Le Marchant, 'An Outline for the Formation of a General Staff to the Army', np, December, 1802. Quoted in Thoumine, op.cit., pp. 102 ff.
60   'Report on Training Officers for the Scientific Corps', 1856, Evidence in reply to Question Four, 'Does any institution for Officers in the Artillery and Engineers, similar to the Senior Department at Sandhurst, appear to you to be desirable?' at pp. 315-74.
61   'Report of the Purchase Commission', 1857, xviii. Evidence of Colonel Lord West at Question 2519.
62   'Report on Training Officers for the Scientific Corps', 1856, p. 286.
63   'The Times', 23 December 1854 and 3 February 1855.
64   The main reports were, 'Report on the State of the Hospitals of the British Army in the Crimea and Scutari', 1855, 'Report to the Rt Hon. Lord Panmure of the Proceedings of the Sanitary Commission Despatched to the Seat of War in the East 1855-1856', 1857, 'First Second and Third Report from the Select Committee on the Army before Sebastopol', 1855 ('The Roebuck Committee'), and 'Report of the Commission of Inquiry into the Supplies of the British Army in the Crimea', 1856.
65   'The Report on the Purchase System', 'Report on Training Officers for the Scientific Corps' and 'The First Report of the Council of Military Education', 1857, 2, xxvii.

CHAPTER 5   THE SEARCH FOR PROFESSIONALISM

1   Correlli Barnett, The Education of Military Elites ,
    p. 20.
2   Masland and Radway, 'Soldiers and Scholars', p. 81.
    Also see Emery Upton, 'Armies of Asia and Europe', New
    York, 1898, and Henry Barnard, 'Military Schools and
    Courses of Instruction in the Science of Art and War in
    France, Prussia, Austria, Russia, Sweden, Switzer-
    land, Sardinia, England and the United States', New
    York, 1872.
3   See, for example, State of the British Army, 'Edinburgh
    Review', December 1886; The War Office, 'Blackwood's
    Magazine', July 1887; Spencer Wilkinson, 'The Brain of
    an Army', London, 1890; Sir William Butler, 'The
    Invasion of England', London, 1882.
4   Sir Harry Verney to Childers, 16 October 1880. Quoted
    in 'The Life of Childers', vol. ii, p. 48.
5   Spencer Wilkinson, 'The Brain of an Army', passim.
6   Robertson, 'From Private to Field Marshal', p. 17.
7   See J.F.C. Fuller, 'The Army in My Time', London, 1936,
    p. 73.
8   T. Wintringham, 'The Story of Weapons and Tactics',
    Boston, 1943, p. 155.
9   J.F.C. Fuller, 'The Conduct of War', London, 1961,
    pp. 121-2.
10 Wolseley to Hartington, 16 April 1885. Sir Frederick
    Maurice and Sir George Arthur, 'The Life of Lord
    Wolseley', London, 1924, pp. 211-15.
11 Edward Thring, 'Education and School', p. 46.
12 See J.F.D. Maurice, 'Learning and Working', London,
    1855. Edited by W.E. Styler, Oxford, 1968, p. 127.
13 'Report from the Select Committee on Scientific
    Instruction', House of Commons, 1867-8, p. 127.
14 'The Times', 20 June 1902.
15 'The Times', 12 June 1902.
16 'The Times', 20 June 1902.
17 'Report of the Royal Commission on Military Education.
    Minutes of Evidence', 1870, XXIV. Evidence of
    Colonel E.G. Hallewell, Commandant of the Royal
    Military College at p. 124.
18 'First Report of the Royal Commission on Military
    Education', 1868-9, XXII, p. 18.
19 Ibid., 'Minutes of Evidence', 1870, XXIV, Evidence of
    Reverend T.A. Southwood, Headmaster, Modern Department,
    Cheltenham College.
20 Ibid., 'Minutes of Evidence', XXIV, p. 230.
21 Ibid., p. 219, Evidence of Reverend Dr Benson, Head-
    master of Wellington College.

22   Ibid., p. 178, Evidence of Major-General H.D. White.
23   A.R. Stanley, 'The Life and Correspondence of Thomas
     Arnold D.D.', London, 1890, p. 198.
24   'Report of the Royal Commission on Military Education,
     Minutes of Evidence', XXIV, p. 186, Evidence of Major-
     General P. Herbert, MP.
25   'Report of the Public Schools Commission', 1864,
     XX, p. 27.
26   Ibid., p. 32.
27   'Return of the Royal Military Academy, Woolwich dated
     30th July, 1868', quoted in 'Parliamentary Papers',
     1870, XXIV (25), Appendix XIII C.
28   'Report of the Military Education Committee', 1902.
     'Minutes of Evidence', (Cd. 983, 1902), p. 7. Evi-
     dence of Mr W.J. Cowthorpe.
29   'Regulations of the Royal Military College', 5 Sept-
     ember, 1869, and 'Regulations for Examination for
     Direct Commission', 1 January 1867, W.O. 32/96, 32/97.
30   'Report of the Royal Commission on Military Education.
     Minutes of Evidence', 1870, XXIV, p. xlvii.
31   Ibid., p. li.
32   Ibid., p. lii.
33   'Report on the Training of Officers for the Scientific
     Corps', 1856, p. xxx.
34   'Report of the Public School Commission', 1864, XX,
     vol. i, pp. 32 and 40. 'Report of the Schools Inquiry
     Commission', 1868, XXXVIII, vol. ii, p. 227.
35   'Report of the Royal Commission on Military Education.
     Minutes of Evidence', 1870, XXIV, p. 223.
36   Ibid., p. 87. Evidence of Canon Heavyside, examiner
     in mathematics to the Council of Military Education.
37   'The Economist', 1 October 1864, at p. 1,225, suggested
     that the lower middle class in England 'were the worst
     educated of any class that is educated at all in this
     country'.
38   Table, calculated from 'Return of the Royal Military
     College', July 1868, at p. 12.
39   'First Report of the Royal Commission on Military
     Education', 1868, XXII, p. 265.
40   Ibid., p. 274.
41   David Newsome, 'A History of Wellington College',
     London, 1959, p. 135.
42   'Report of the Royal Commission on Secondary Education',
     1895, (c. 7862, 1895).
43   'Final Report of Departmental Committee on Royal
     College of Science', 1906, Questions 38, and 326-33.
     Also see 'Journal of the Iron and Steel Institutes',
     1906, pp. 262-8.

44 'Report of the Royal Commission on Military Education.
    Minutes of Evidence', 1870, XXIV. Memorandum by
    Colonel Addison, Superintendant of Studies, Sandhurst,
    23 March 1869.
45 For a further exposition of this point, see Flann
    Campbell, Latin and the Elite Tradition in Education,
    'British Journal of Sociology', vol. 19, no. 3,
    September 1968, p. 309.
46 R.R. Bolger, 'The Classical Inheritance and its
    Beneficiaries', Cambridge, 1954, p. 1.
47 'Report of the Royal Commission on Military Education.
    Minutes of Evidence', 1870, XXIV, p. 22.
48 See 'Report of the President of Queens' College,
    Belfast', 1868, XXII, pp. 317-417.
49 'Report of the Royal Commission on Military Education.
    Minutes of Evidence', 1870, XXIV, p. 334. Evidence
    of Captain C.B. Brackenbury RA.
50 For a further discussion of these points, see G.
    Harries-Jenkins, Dysfunctional Consequences of Military
    Professionalization, in Janovitz and van Doorn (eds),
    'On Military Ideology', pp. 141-65.
51 'Report of the Royal Commission on Military Education.
    Minutes of Evidence', 1870, XXIV, p. 336.
52 Ibid., p. 468. Evidence of Professor J. Sylvester
    FRS, Professor of Mathematics at the Academy.
53 Ibid., p. 475. Memorandum by Colonel G.T. Field,
    Inspector of Studies at the Academy.
54 C. Erickson, 'British Industrialists. Steel and
    Hosiery: 1850-1950', London, 1959, p. 58.
55 'Report of the Military Education Committee, 1902'.
    Memorandum by Lieutenant-Colonel A.M. Murray, Assistant
    Commandant of the Royal Military Academy, 1 July 1901,
    at pp. 105-7.
56 'Report of the Royal Commission on Military Education.
    Minutes of Evidence', 1870, XXIV, Evidence of Profes-
    sor J. Sylvester. The proposals to unite the two
    colleges in 1858 are reviewed in W.O. 33/6, p. 793.
57 Ibid., Evidence of Captain T.B. Strange, RA.
58 Sir Charles Dilke, The Present Position of European
    Politics or Europe in 1887, 'Fortnightly Review', 1887,
    p. 337.
59 Sir Frederick Maurice, The Balance of Military Power
    in Europe, 'Blackwood's Magazine', July 1887, p. 147.
60 See 'Second Report from the Select Committee on Army
    and Navy Estimates', 1887, VIII, no. 3668.
61 'Hansard', vol. 9, 9 March 1893, p. 1,525.
62 The Condition of the Army, 'Spectator', 20 November
    1897, p. 726.
63 'The Times', 7 August 1902.

64 Calculated from evidence published in 'Report of the Military Education Committee', 1902, passim.
65 See, for example, 'The Times', 12 and 20 June 1902.
66 'Report of the Military Education Committee', 1902, p. 214. Return by Major-General A.E. Turner, Inspector-General of Auxiliary Forces.
67 Ibid., Evidence of the Duke of Bedford at p. 226.
68 Ibid.
69 Ibid., Evidence of Lord Raglan at p. 226.
70 Ibid., Appendix ix, p. 65. Memorandum by A.A. Somerville, Eton College.
71 Ibid., p. 58. Memorandum by J.S. Phillips, Headmaster of Bedford Grammar School.
72 Ibid.
73 M.E. Sadler, The Unrest in Secondary Education in Germany and Elsewhere, in 'Special Reports on Educational Subjects', vol. 9 (1902), p. 50.
74 'Report of the Military Education Committee', 1902, Appendix vii, p. 60, Memorandum on 'Examinations for Entrance to the Army', by Major W.H. James, RE.
75 Cited in John Terraine, 'Douglas Haig, the Educated Soldier', London, 1963, p. xiii.
76 'Report of the Military Education Committee', 1902, p. 152. Evidence of Lieutenant-Colonel Bourne.
77 'The Times', 12 June 1902.
78 Terraine, op.cit., p. 1.
79 'The Times', 12 June 1902.
80 'Report of the Military Education Committee, 1902, p. 29.
81 Ibid., p. 30.
82 'Report of the War Office (Reconstitution) Committee (Part II)', (Cd. 1968), 1904, VIII, p. 22.
83 Willoughby Verner, 'The Military Life of the Duke of Cambridge', London, 1905, vol. i, p. 407. Duke of Cambridge to Cardwell, 21 November 1869.
84 'Report of the Royal Commission Appointed to Enquire into the Civil and Professional Administration of the Naval and Military Departments', (Cd. 5979), 1890, XIX, pp. xxix-xxx.
85 Hamer, 'The British Army, Civil-Military Relations 1885-1905', p. 36.
86 The General Staff was established in 1905 to meet the recommendations of the 'Report of the War Office (Reconstitution) Committee' (the Esher Committee).
87 This criticism was repeated in 'The Times' over a period of thirty-five years. See 'The Times', 18 January 1867 and 12 June 1902.
88 Robertson, op.cit., p. 78.
89 Brian Gardner, 'Allenby', London, 1965, p. 14.

90 Quoted by Goodwin-Austen, 'The Staff and the Staff
    College', p. 190. See also evidence of Colonel
    Clifford to the Military Education Commission, 1868.
    'Report of the Royal Commission on Military Education.
    Minutes of Evidence', 1870, XXIV, p. cxvii.
91 Newsome, 'A History of Wellington College', p. 61.
    'Gentlemanideal' was a portmanteau word coined by
    the German writer Dibelius, to describe the English
    veneration for the specific life-style and its asso-
    ciated attitudes which were adopted by the con-
    temporary élite.
92 Compton, 'Open Competition and the Indian Civil
    Service, 1850-1876', p. 270.
93 'Report of the Royal Commission on Military Education.
    Minutes of Evidence', 1870, XXIV, p. 344.
94 Ibid., p. 355.
95 Ibid.
96 See K. Ekirich, 'The Civilian and the Military', New
    York, 1956.
97 E. Root, 'Five Years of the War Department', Washing-
    ton, 1904, pp. 58 and 226.
98 'Report of the Royal Commission on Military Education.
    Minutes of Evidence', 1870, XXIV. Evidence of
    Colonel J.W. Armstrong, at p. 365.
99 Ibid. War Office Returns dated 1 January 1868.
100 Quoted in Godwin-Austen, op.cit., p. 207.
101 Ibid., p. 235.
102 Ibid., p. 231.
103 Robertson, op.cit., p. 175.
104 'Report of the Royal Commission on the War in South
    Africa. Minutes of Evidence', 1904, XL, Evidence
    of Mr L.S. Amery.
105 Cairnes, 'Social Life in the British Army', p. 175.
106 'Report of the Royal Commission on the War in South
    Africa', 1904, XL, p. 53.
107 Ibid.
108 'Report of the Committee Appointed to consider the
    Education and Training of the Officers of the Army',
    1903, p. 21.
109 Ibid.
110 Ibid., p. 29.
111 'Report of the Royal Commission on the War in South
    Africa', 1904, XL, p. 441.
112 Viscount Esher to Lord Knollys, 27 May 1906. M.V.
    Brett (ed.), 'Journals and Letters of Viscount Esher',
    London, 1934-8, vol. ii, p. 166. (Hereafter referred
    to as 'Esher Journals'.)
113 'Report of the Committee appointed to consider the
    Education and Training of the Officers of the Army',
    1902, p. 2.

CHAPTER 6   THE TASK OF THE ARMY

1   'Report of the Royal Commission on Warlike Stores', 1887, Evidence of Lord Wolseley at paras 2641-3.
2   The controversy over the conscription issue at the end of the Victorian period is discussed in W. Ropp, Conscription in Great Britain, 'Military Affairs', vol. xx, 1956, pp. 71-6. See in addition Colonel J.K. Dunlop, 'The Development of the British Army 1899-1914', London, 1938, and David James, 'The Life of Lord Roberts', London, 1954.
3   General Sir Edward Bruce Hamley, 'The Operations of War', (revised edition by Major General Sir George Astor), Edinburgh, 1923, p. 56.
4   Hamley, op.cit., p. 431.
5   Steevens, 'With Kitchener to Khartum', pp. 286-7.
6   W.S. Churchill, 'The River War', London, 1899, vol. ii, pp. 282-3.
7   Ibid., p. 287.
8   Steevens, op.cit., p. 293.
9   Ibid., p. 287. A 'zariba' was a hedged or palisaded enclosure.
10 Cyril Falls, The Reconquest of the Sudan, 1896-9, in Brian Bond (ed.), 'Victorian Military Campaigns', p. 296.
11 Hamley, op.cit., p. 432.
12 See Sir Ian Hamilton, 'The Fighting of the Future', London, 1885, and Brian Bond, Doctrine and Training in the British Cavalry 1870-1914, in Michael Howard (ed.), 'The Theory and Practice of War', London, 1965.
13 'Life of Childers', vol. ii, pp. 25-7.
14 See E. Holt, 'The Strangest War', London, 1962, pp. 196-7.
15 Janowitz, 'The Professional Soldier', pp. 418 ff.
16 Lord Elton, 'General Gordon', London, 1954, p. 53.
17 See Major T.J. Holland and Captain H.M. Hozier, 'Record of the Expedition to Abyssinia, compiled by order of the Secretary of State for War', London, 1870, vol. i, pp. 235-6.
18 See Captain Henry Brackenbury, 'The Ashanti War', Edinburgh, 1874, vol. i, p. 184.
19 Holt, op.cit., pp. 148 and 162.
20 Hamer, 'The British Army, Civil-Military Relations 1885-1905', p. 59.
21 'Second Report from the Select Committee on Army and Navy Estimates', 1887, VIII, nos 3571-81.
22 Robertson, 'From Private to Field Marshal', p. 130.
23 'Appendices to the Minutes of Evidence before the Royal Commission on the War in South Africa', 1904, XLII, p. 282.

24   'Report of the Royal Commission on the War in South
     Africa', 1904, XI, pp. 10-13.
25   Quoted in Spenser Wilkinson, The Cabinet and the War
     Office, 'Quarterly Review', October 1903, p. 590.
26   See Sir George S. Robertson, 'The Story of a Minor
     Siege', London, 1898, and Captain G.J. and Captain
     Frank E. Younghusband, 'The Relief of Chitral'.
27   Sir Charles Dilke and Spenser Wilkinson, 'Imperial
     Defence', London, 1897, p. 144.
28   Edward Gillett, 'A History of Grimsby', Oxford, 1970,
     p. 289.
29   See John Saville, Unions and Free Labour: Background
     to the Taff Vale Decision, in Asa Briggs and John
     Saville (eds), 'Essays in Labour History', London,
     1960.
30   W.C.B. Tunstall, Imperial Defence 1815-1870, in 'The
     Cambridge History of the British Empire', Cambridge,
     1940, vol. ii, p. 808.
31   Southgate, 'The Passing of the Whigs', p. 169.
32   See Jagnishan Mahajan, 'The Annexation of the Punjab',
     Karachi, 1949.
33   See Fanja Banjee, 'The Military System of the Sikhs,
     1799-1849', Delhi, 1964.
34   Captain C.E. Calwell, 'Small Wars: Their Principles
     and Practice', London, 1896.
35   Hamley, op.cit., pp. 87-102, at p. 99.
36   See Bernard Ash, 'The Lost Dictator', London, 1968,
     pp. 25-7.
37   Hamley, op.cit., p. 346.
38   Correlli Barnett, 'Britain and Her Army', London,
     1970, p. 328.
39   Lieutenant-Colonel H. de Watteville, 'Lord Roberts',
     Glasgow, 1938, pp. 107-8.
40   Steevens, op.cit., pp. 199-200.
41   Steevens, op.cit., pp. 73-4.
42   Hamley, op.cit., p. 66.
43   Ibid., p. 420.
44   Correlli Barnett, op.cit., p. 319.
45   Hamley, op.cit., pp. 59-60.
46   See Lord Edward Gliechen, 'A Guardsman's Memories',
     London, 1932. These volunteer officers demonstrated
     clearly the relationship between the landed interest
     and the officer corps, and the way in which the former
     were quick to participate in duties which were 'out of
     the ordinary'. Apart from Count Gliechen, other
     officers included Captain Lord St Vincent, killed at
     the Atbara, Captain Lord Cochrane, Lieutenant Lord
     Rodney, Lieutenant Lord Binning, and a Marines offi-
     cer, Lieutenant Charles Townshend.

47   Hamley, op.cit., p. 416.
48   Steevens, op.cit., pp. 294–5.
49   Correlli Barnett, op.cit., p. 340.
50   Pratt, 'Precis of Modern Tactics', quoted in Hamley, op.cit., at p. 425.
51   'Cavalry Manual for 1907'.
52   Introduction by Major-General J.F.C. Fuller, referring to Haig's obsession with the use and advantage of cavalry, in L. Wolff, 'In Flanders Field', London, 1959, at p. 13.
53   Hamley, op.cit., p. 426.
54   Steevens, op.cit., p. 56.
55   General Sir Ian Hamilton, 'Listening for the Drums', London, 1944, p. 172.
56   Colonel J.F. Maurice, 'The Military History of the Campaign of 1882 in Egypt', London, 1887, Appendix 1.
57   Memorandum of the Duke of Cambridge, December 1852, 'Observations on the Organization of the British Army at Home', quoted in G.R. St Aubyn, 'The Royal George', p. 58.
58   Gardner, 'Allenby', p. 65. Apart from the Household Cavalry, who were based on London and Windsor, the remainder of the cavalry in England and Ireland were located in eleven depots. These ranged from York (2nd Dragoons (Royal Scots Greys)) to Canterbury (6th Dragoon Guards (Carabiners)) and from Dublin (5th (Royal Irish) Lancers) to Norwich (12th Prince of Wales Royal Lancers). See R. Money Barnes, 'The British Army of 1914', London, 1968.
59   Captain Owen Wheeler, 'The War Office Past and Present', p. 232.
60   The State of the British Army, 'Edinburgh Review', January 1885, pp. 211–12.
61   'Report of the Select Committee on Finance', 1828, Evidence of the Duke of Wellington.
62   See John Prebble, 'The Highland Clearances', London, 1963.
63   P.G. Rogers, 'Battle in Bossendon Wood. The Strange Story of Sir William Courtenay', London, 1961.
64   M. Hovell, 'The Chartist Movement', London, 1918, p. 290. George Rude, 'The Crowd in History, 1730–1848', London, 1964, pp. 259 ff.
65   John Prebble, op.cit., pp. 180–5.
66   See Sir John Fortescue, The Army, in G.M. Young (ed.), 'Early Victorian England', London, 1934, vol. i, p. 352.
67   'Report of the Commission on Naval and Military Promotion', 1838, Evidence of Lieutenant-General Sir William Gordon, Quartermaster-General, Question 1843 at p. 113.

68   Biddulph, 'Lord Cardwell at the War Office', p. 40.
69   'Punch or the London Charivari', vol. XLV, 3 October
     1863, pp. 138-9.
70   Quoted in Brian Bond, Prelude to the Cardwell
     Reforms, 'Journal of the Royal United Service Insti-
     tution', 1961, p. 233.
71   John Prebble, op.cit., pp. 295-302.
72   'Life of Childers', vol. ii, pp. 192-3.
73   'Letters to Queen Victoria', 2nd series, vol. iii,
     pp. 640-2. Roseberry to Ponsonby, 24 April 1885.
74   J.A. Spender, 'The Life of Sir Henry Campbell-
     Bannerman', London, 1922, vol. i, p. 126.
75   By the Government of India Act, 1858, Indian revenues
     could not be used to support military operations
     outside India without the consent of Parliament,
     except in urgent necessity or to repel invasion. Even
     so, Indian troops were employed outside the sub-
     continent in Malta in 1878, in Egypt in 1882, in the
     Sudan in 1886, and in China in 1900. There were
     Indian garrisons in Ceylon, Mauritius, Singapore and
     China. See 'The Cambridge History of the British
     Empire', vol. v, p. 210.
76   Dilke and Wilkinson, op.cit., p. 138.
77   Ibid., p. 158.
78   G.F.R. Henderson, 'The Science of War', London, 1910,
     p. 405.
79   Dilke and Wilkinson, op.cit., p. 145.
80   See J.K. Dunlop, 'The Development of the British Army
     1899-1914', Appendix A, p. 307.
81   Geoffrey Cousins, 'The Defenders', London, 1968,
     p. 158.
82   Viscount Esher to A.J. Balfour, 15 December 1903.
     'Esher Journals', vol. ii, p. 33.
83   Quoted in A.J. Marder, 'The Anatomy of British Sea
     Power', New York, 1940, p. 65.
84   Sir John Fisher to Viscount Esher, 15 March 1909,
     'Esher Journals', vol. ii, p. 375.
85   Esher to Sir John Fisher, 21 February 1905, 'Esher
     Journals', vol. ii, p. 75. The quotation is taken
     from a paper drawn up by Viscount Esher for the Prime
     Minister.
86   Steevens, op.cit., p. 23.
87   Girouard, a French Canadian, was subsequently appointed
     whilst still a subaltern in the Engineers, Director-
     General of all the Egyptian railways.
88   Robert Swinhoe, 'Narrative of the North China Campaign
     of 1860', London, 1861, p. 144.
89   See Phillip Magnus, 'Kitchener: Portrait of an Imper-
     ialist', London, 1958, pp. 81-2.

90 Ibid., p. 122.
91 Brian Bond, The South African War, 1880-1, in 'Victor-
    ian Military Campaigns', p. 215.
92 General Sir Hope Grant (ed. H. Knollys), 'Incidents
    in the China War of 1860', Edinburgh, 1875, p. 142.
93 H.D. Napier, 'Field Marshal Lord Napier of Magdala',
    London, 1927, p. 204.
94 For a further comment on this point, see M.D. King,
    Science and the Professional Dilemma, in Julius Gould
    (ed.), 'Penguin Social Sciences Survey, 1968', London,
    1968.
95 S.E. Finer, 'The Man on Horseback', London, 1964,
    p. 25.
96 Despatch by Lord Curzon to St John Brodrick, Secretary
    of State, 23 March 1905. Quoted in Magnus, 'Kitchener:
    Portrait of an Imperialist', p. 216.

CHAPTER 7   THE ARMY AND ITS POLITICAL ATTITUDES

1 Harold Laswell, 'Politics. Who Gets What, When and
   How', New York, 1936.
2 Sources: for 1853, Charles R. Dod, 'The Parliamentary
   Companion for 1853', London, 1853, and 'Hart's Army
   List for 1853'. For 1885, 'The Constitutional Year
   Book, 1885', London, 1885, and for 1898, 'The Constitutional Year Book, 1898', London, 1898 together with
   the Army List for 1885 and 1898.
3 Guttsman, 'The British Political Elite', p. 111.
4 Sir Lewis Napier, 'The Structure of Politics at the
   Accession of George III', London, 1965, pp. 24-8,
   points out that at the General Election of 1761, sixty-
   four army officers, actually in the service, were
   elected. These included the best-known members of the
   military élite: Field Marshal Lord Lignier who was
   the Commander-in-Chief, Lord Granby, Sir John Mordaunt,
   J.S. Conway, Robert Clive, George Townsend, John
   Burgoyne, William Howe and Charles Cornwallis. Napier
   stresses that his calculation does not include officers
   who had sold out, or who had been dismissed from the
   army, or men who had served in their youth or in time
   of emergency.
5 G.P. Judd, 'Members of Parliament, 1734-1832', New
   Haven, 1955.
6 See Bernard Cracroft, The Analysis of the House of
   Commons, or Indirect Representation, in 'Essays on
   Reform', London, 1867, pp. 156-65 and 327-9 (statistical tables).

7   Speech of Lord Chesterfield in the House of Lords,
    3 February 1741, 'Journals of the House of Lords',
    vol. xxv, p. 586.
8   The Duke of Buckingham, 'Memoirs of the Courts of
    William IV and Victoria', London, 1861, vol. ii,
    p. 287. Londonderry was perhaps prejudiced since when
    he was the heir to the title, he had enjoyed £1,500 a
    year as one of the eleven commissioners for the affairs
    of India, 'very snug and profitable places'. See 'The
    Black Book', London, 1820, p. 12.
9   Bonham to Peel, 29 September 1845, quoted in Norman
    Gash, 'Politics in the Age of Peel', London, 1953,
    p. 379.
10 Edward Baines, 'History, Directory and Gazetteer of
    the County of York', London, 1823, vol. ii, p. 513.
11 Donald Southgate, 'The Passing of the Whigs, 1832-
    1880', points out that the Cowper family had sixty-
    seven years' service as MPs, a total rarely exceeded
    during this period by other families.
12 'Parliamentary Papers', 1833, xi, pp. 4-73.
13 L. Strackey and R. Fulford (eds), 'Greville Memoirs',
    London, 1938, vol. iii, p. 340.
14 'Parliamentary Papers', 1836, xix, p. 77. Chetwynd
    was the second son of Sir George Chetwynd of Brocton
    Hall, a small estate four miles south-east of Stafford.
15 The intricacies of Victorian family networks and their
    effect on the political attitudes of officers was seen
    again in the military and public career of the 3rd
    Earl of Lonsdale's brother-in-law, Rt Hon. Gerard
    James Noel (1823-1911). The son of the 1st Earl of
    Gainsborough, Noel was a retired captain of the 11th
    Hussars who, when MP for Rutland 1847-83, was Parlia-
    mentary Secretary to the Treasury, Chief Commissioner
    of Works and Public Buildings and a Privy Councillor.
16 'Parliamentary Papers', 1835, xvii, pp. 245-448.
    Thirty-five rounds of ammunition were returned as
    having been used, the casualties totalling three — a
    boy of seventeen shot through a leg which afterwards
    had to be amputated, a boy of eleven shot through an
    ankle, and a boy of fifteen shot through the heel.
17 Southgate, op.cit., p. 95.
18 Quoted in 'Greville Memoirs', vol. i, at p. 26.
19 R.W. Jeffrey, (ed.) 'Dyott's Diary', London, 1907,
    vol. ii, p. 258.
20 Ibid., vol. ii, p. 343.
21 See J.H. Hanham, 'Elections and Party Management',
    London, 1959, p. 251.
22 Lord George Hamilton, 'Parliamentary Reminiscences
    and Reflections 1886-1906', London, 1922, p. 15.

23 See W.W. Bean, 'The Parliamentary Representation of
    the Six Northern Counties of England', Hull, 1890.
24 James Grant, 'Random Recollections of the Lords and
    Commons', 2nd series, London, 1838, vol. ii, pp. 66 ff.
25 Gash, op.cit., p. 455.
26 See Biddulph, 'Lord Cardwell at the War Office', p. 152.
27 Cardwell to Gladstone, 20 November 1870. Cited in
    Biddulph, op.cit., p. 102.
28 Gladstone to Cardwell, 8 July 1871. Quoted in John
    Morley, 'Life of Gladstone', London, 1905, vol. ii,
    p. 649.
29 E. Peel (ed.), 'Recollections of Lady Georgiana Peel',
    London, 1920, p. 209.
30 Mackail and Wyndham, 'Life and Letters of George
    Wyndham', p. 67.
31 Gash, op.cit., p. 336.
32 Guttsman, op.cit., p. 142.
33 Richard William Penn Curzon Howe, 3rd Earl Howe, was
    another example of the links which were established
    between these families. His mother was the sister of
    the Earl of Cardigan, his wife was the daughter of
    General George Anson and his sister had married the
    Duke of Beaufort, who was Lord Raglan's great-nephew.
34 S. and B. Webb, 'The Parish and the County', London,
    1906, p. 286.
35 Earl of Rosslyn to Sir Duncan McDougall, 13 May 1859.
    Quoted in Geoffrey Cousins, 'The Defenders', p. 102.
36 F. Leary, 'History of Earl of Chester's Regiment of
    Yeomanry Cavalry, 1798-1897', Edinburgh, 1898.
37 'The Times', 24 May 1859.
38 'Hansard', 4th Series, vol. 9, 14,444.
39 E. Burke, 'Correspondence of Edmund Burke', London,
    1844, vol. i, p. 141.
40 Speech by Brougham in the House of Commons 7 February
    1828, 'Hansard', vol. xviii.
41 This qualification had been raised in 1744-5 (18 Geo.
    II c. 20) to ownership of land within the county worth
    £100 a year or £300 a year in reversion, except for
    peers, their eldest sons and heirs and the eldest sons
    and heirs of persons owning £600 a year in land.
42 Speech by the Earl of Albemarle in the House of Lords.
    'Hansard', 1875, vol. 223, pp. 765-70.
43 Thompson, 'English Landed Society in the Nineteenth
    Century', p. 128.
44 T.H.S. Escott, 'Social Transformation of the Victorian
    Age', London, 1897, pp. 98-9.
45 J.N. Lee, 'Social Leaders and Public Persons', Oxford,
    1963, pp. 55-6.

46 'Hansard', 1888, vol. 329, pp. 918-33.
47 F.W. Maitland, The Shallows and Silences of Real Life, 'The Reflector', February 1888.
48 For the opinion of the Judge Advocate General in 1880 on the conflict of duties which arose when an officer was concomitantly an MP, see W.O. 33/32, p. 820 and W.O. 83/4, pp. 702-5.
49 Speaker Brand to Childers, 17 December 1880. 'Life of Childers', vol. i, p. 282.
50 Guttsman, op.cit., p. 82.
51 Bernard Holland, 'The Life of Spencer Compton, 8th Duke of Devonshire', London, 1911, vol. i, pp. 402-3.
52 'Fortnightly', vol. ccxxvi, October 1885, Three Programmes.
53 Salisbury to Queen Victoria, 24 September 1886. 'Letters of Queen Victoria', 3rd series, 1886-1901, vol. i, pp. 194-5.
54 'Hansard', vol. 326, 14 May 1888, pp. 91-7, 100-1.
55 'Report of the Royal Commission appointed to inquire into the system under which patterns of warlike stores are adopted, and the stores obtained and passed for her Majesty's service, with appendices', (Cd. 5062-1, 1887) XV, (the Stephen Commission); 'First Report of the Royal Commission appointed to inquire into the civil establishments of the different offices of state at home and abroad, with minutes of evidence, appendix', (Cd. 5-26, 1887), XIX, (the Ridley Commission); 'Preliminary and further reports (with appendices) of the Royal Commissioners appointed to enquire into the civil and professional administration of the naval and military departments and the relation of those departments to each other and to the Treasury', (Cd. 5979, 1890), XIX, (the Hartington Commission).
56 'Report of the Royal Commission on Warlike Stores', 1887, p. x.
57 Wolseley to Brodrick, 3 October 1890. Quoted in Hamer, 'The British Army, Civil-Military Relations, 1885-1905', p. 219.
58 'Report of the committee appointed to inquire into the organization and administration of the manufacturing departments of the army; with minutes of evidence, appendix and index', (Cd. 5116, 1887), XIV, (the Morley Committee), pp. xiv ff.
59 See J.R. Seeley, 'The Expansion of England', London, 1883, p. 154.
60 'First Report of the Royal Commission on Civil Establishments', 1887, XIX, p. vi.
61 See Brian Bond, The Retirement of the Duke of Cambridge, 'Journal of the Royal United Service Institution', 1961, p. 544.

62 'Hansard', vol. 327, 29 June 1888, p. 1697.
63 'Reports of the Royal Commission on the Naval and
    Military Departments', 1890, XIX, p. 3. The members
    of the Commission were Lord Hartington, Lord Randolph
    Churchill, W.H. Smith, Sir Henry Campbell-Bannerman,
    Sir Richard Temple, Admiral Richards, General Sir
    Henry Brackenbury and Mr Ismay. The Secretary was Sir
    George Clarke, an officer of the Royal Engineers who
    had served with Wolseley in Egypt.
64 The opposition of Queen Victoria to the Report of the
    Hartington Commission can be seen in 'Letters of Queen
    Victoria', 3rd series, vol. i, particularly the Queen
    to Sir Henry Ponsonby, 20 March 1890 at pp. 582-3.
65 'Hansard', vol. 333, 11 March 1889, p. 1,478.
66 'Hansard', vol. 346, 26 June 1890, pp. 101-2.
67 'Hansard', vol. 22, 16 March 1894, p. 465.
68 'Hansard', vol. 31, 15 March 1895, p. 1,204.
69 'Hansard', vol. 54, 25 February 1895, p. 116.
70 Spender, 'Life of Sir Henry Campbell-Bannerman',
    vol. i, p. 147. Sir Ralph Thompson to Campbell-
    Bannerman, 30 May 1895.
71 'Hansard', vol. 34, 21 June 1895, pp. 1,673-8. The
    discussions which took place are analysed in Hamer,
    op.cit., pp. 160-1; Wheeler, 'The War Office Past
    and Present', pp. 246-8 and Brian Bond, The Retirement
    of the Duke of Cambridge.
72 S. Gwynne and G.M. Tuckwell, 'Life of Sir Charles
    Dilke', London, 1917, vol. ii, p. 424. 'Hansard',
    vol. 36, 19 August 1895, pp. 245-6.
73 'Order-in-Council, 21st November 1895', 1896; 'Memoran-
    dum showing the Duties of the Principal Officers and
    Departments of the War Office Under the Order-in-
    Council dated 21st November 1895',(Cd. 7987, 1896) LI.
74 Writing in 1904, the editor of 'Review of Reviews'
    suggested that because of the Esher Committee, 'The
    Gordian knot which has baffled generations of reform-
    ers has at last been cut thanks to the determined
    action of three trusted and competent men who stood
    outside the Parliamentary chaos'. 'Review of
    Reviews', 10 March 1903, p. 214.
75 Guttsman, op.cit., p. 168.
76 'Memorandum by Field-Marshal Viscount Wolseley ad-
    dressed to the Marquis of Salisbury relative to the
    working of the Order in Council of 21st November 1895',
    (Cd. 512, 1901) XXXIX, pp. 3-4.
77 'Hansard', vol. 90, 5 March 1901, p. 545.
78 Bernard Ashe, 'The Lost Dictator', p. 52.
79 Quoted in J. Wheeler-Bennett, 'The Nemesis of Power',
    London, 1953, pp. 108-9.

80   A.P. Ryan, 'Mutiny at the Curragh', London, 1956,
     p. 142.
81   'Report of the Royal Commission on the War in South
     Africa, Minutes of Evidence', 1904, XL, para. 13,621.
     Butler's justification for the action he took in
     South Africa on the grounds that he was not the ser-
     vant of Rhodes's South Africa League, must be seen,
     however, against his personal feelings which led him,
     throughout his life, to support the cause of subject
     peoples.
82   'Parliamentary Papers: Egypt No. 1', London, 1914,
     Cd. 1358.
83   'Who Was Who, 1897–1916', p. 284.
84   'Royal Archives', George V F674/17. Quoted in Sir J.
     Fergusson, 'The Curragh Incident', London, 1964.
85   Magnus, 'Kitchener: Portrait of an Imperialist',
     p. 299.
86   D. Lloyd George, 'War Memoirs of David Lloyd George',
     London, 1938, vol. i, p. 83.
87   'War Memoirs of David Lloyd George', vol. ii,
     pp. 1,668–9.

CHAPTER 8   POSTSCRIPT

 1   'The Times History of the War in South Africa, 1899–
     1902', London, 1900–9, vol. ii, p. 40.
 2   'Report of the Royal Commission on the War in South
     Africa', 1904, XL, p. 22.
 3   Lord Wolseley's speech in the House of Lords. 'Hansard',
     vol. 90, 4 March 1901, pp. 327–43. Campbell-Bannerman,
     as leader of the Liberal Party, equally sought to
     blame the Conservative government of the day, arguing
     that he had never refused the military anything, guns
     or stores of any sort for which they had asked.
     (Spender, 'Life of Sir Henry Campbell-Bannerman',
     vol. 1, p. 273).
 4   'Report of the Royal Commission on the War in South
     Africa', 1904, IL, pp. 44–5.
 5   Correlli Barnett, 'The Collapse of British Power',
     London, 1972, p. 43.
 6   Memorandum of the First Sea Lord, March 1906. Quoted
     in A.J. Marder, 'From the Dreadnought to Scapa Flow:
     The Royal Navy in the Fisher Era', vol. i, 'The Road
     to War 1904–14', Oxford, 1961, p. 31.
 7   Correlli Barnett, 'The Collapse of British Power',
     p. 43.
 8   Anthony Brett-James, 'The British Soldier in the
     Napoleonic Wars, 1793–1815', London, 1970, p. 44.

9   See Esme Wingfield-Stratford, 'The Squire and his
    Relations', London, 1956, pp. 251-6.
10 Correlli Barnett, 'The Swordbearers: Studies in
    Supreme Command in the First World War', London,
    1966, p. 208.
11 See H.J. Mackinder, The Geographical Pivot of History,
    'Geographical Journal', vol. 23, 1904, pp. 421-44.
    For a criticism of the Mackinder hypothesis see J. de
    Blij, 'Systematic Political Geography', New York,
    1967, pp. 130-7 and A.R. Hall, Mackinder and the
    Course of Events, 'Annals of the American Association
    of Geographers', vol. 65, 1955, pp. 109-26.
12 'Geographical Journal', vol. 23, 1904, p. 438.
13 David Walder, 'The Short Inglorious War', London,
    1973, p. 19.

Index

Militarism, 7, 8, 265
Military casualties, 1,
   174, 189
Military crammers, 144-5,
   147, 156
Military Education Com-
   mittee (1901), 156-7
Military strategy, 185ff
Military tactics, 195-6,
   200
Military virtues, 58, 105
Military weapons, 192-4,
   263
Minto, Earl of, 35
Mitchell, General John,
   107-8
Montgomery-Cunningham, Sir
   William, 34
Moore, Isaac, 46
Morley, Samuel, 18
Mosca, Gaetano, 9, 13
Mott, Edward, 47-8
Muncaster, Lord, 33
Munro of Ross, Sir Charles,
   31
Murray of Ochtertyre,
   General Sir George, 245

Napier, Sir Charles, 251
Napier, General Sir
   Robert (later Lord),
   177, 191, 192-3, 200,
   210, 213, 219
Napier, Sir William, 107,
   109
Neville, Hugh, 46
Newdigate family, 254
New Zealand War, Second
   (1860-1), 178
New Zealand War, Third
   (1863-6), 176
Norcliffe, Cecil, 37
Northbrook Committee, 243

O'Brien, Sir John, 66
O'Connor, Sir Luke, 47
Officers: administrative
   appointments, 35, 37,
   41, 84, 104, 177,

217-18, 248ff; amateurism,
17-18, 60, 100, 102, 212,
216, 266, 279; authority
of, 16, 38, 104, 276;
commercial interests,
260-1; conservatism, 48,
56, 72, 93, 193, 212,
276-7; education, 17, 68,
75, 95, 103, 105, 138,
277; élitism, 49-50, 54,
91, 95, 100, 104, 122,
165, 197, 247; evaluation
of merit, 82-3, 94, 162,
212, 277; expenses of
life-style, 86-8, 95,
97-8; flâneurs, as, 19,
212, 216, 276; half pay,
71-3, 89-91; individual-
ism, 198-9; intellectual
ability, 104, 109, 115,
117-21, 124, 134, 141-2,
149, 154, 156-7, 166-9,
277; junior ministers, as,
255, 257-8; local admini-
stration, part of, 255,
257-8; loyalty to Crown,
13, 19; magistrates, as,
252-7; membership of
clubs, 9, 35-6; Members
of Parliament, as, 23,
28, 33, 34, 36-7, 56,
220ff, 258-9; mess life,
96, 100, 105; neo-feudal
attitudes, 4, 17, 56, 61,
95, 129, 212, 251; pay,
85-8, 101; pensions, 85,
88-9; political attitudes,
9, 19, 47, 216, 224ff,
243, 261-2, 270-3; poli-
tical patronage, 246-7;
professionalism, 3, 6, 12,
15, 18, 60, 91-2, 94, 96,
101, 108-9, 114, 116, 120,
143-4, 166-8, 172, 191,
211, 214-16, 271, 279;
promotion of, 70-1, 76,
78-80, 90, 125; purchase
of commissions, 14, 18,
59ff, 93-4, 135, 275;

qualities of officer-
ship, 17, 52, 81, 99,
104-5, 136, 169, 197,
277-9; relationship
with baronetage, 41-3;
relationship with
landed interest, 9,
21, 25-38, 43, 47, 52-
3, 55, 84, 245-6, 251,
253, 259; relationship
with middle class, 4,
44-6, 56, 60, 91-2,
100, 169, 270-3; rela-
tionship with peerage,
39-41; relationship
with soldiers, 3, 16,
52-3, 98; rural iden-
tification, 249-50,
259; selection of, 12,
16, 20-2, 24, 44-5,
56, 92, 98, 105, 124,
152, 154-5, 169; self-
recruitment, 66;
social and public
responsibility, 18,
36, 37, 250-1; social
background, 3, 16, 43-
5, 52, 95, 247; social
values, 3, 5, 10, 12,
15-16, 37, 43, 48, 52-
4, 56, 95, 99, 104, 160,
218, 247, 278; sociali-
zation of, 5, 29, 96-7,
103, 212, 219, 245-6,
251, 277; study of
theory, 105-9, 190
Open competition, 13-14,
20

Otway, Arthur J., 207,
261

Paget family (Marquess of
Anglesey), 40, 50, 228,
234-6, 249, 271
Pakenham, E.W., 228
Palmerston, Lord, 84, 204
Panmure, Lord, 86
Parker, Sir William, 33

Parliamentary constituencies:
Co. Antrim, 228; Ayrshire,
246; Barnstaple, 31; Beaumaris, 228; Berwick, 56,
244; Beverley, 36; Buck-
ingham, 227; Cambridge-
shire, 228; Chatham, 229,
243; Cheltenham, 23;
Cheshire, East, 36; Che-
shire, Mid, 251; Chiches-
ter, 222, 233; Chippenham,
233; Cirencester, 228; Co.
Clare, 226; Cockermouth,
229, 241; Cumberland East,
233; Cumberland West, 227,
233, 241; Devonport, 226,
242; Dungannon, 227;
Durham North, 34, 239;
Durham South, 239, 241;
Eye, 230; Co. Fermanagh,
232; Forfar, 243; Frome,
222; Gloucestershire,
East, 249; Gloucester-
shire, West, 227; Had-
dington, 225; Hastings,
244; Herefordshire, 227;
Hertford, 230; Hunting-
don, 228; Kincardine,
226; Lancashire, North,
56, 240; Lancashire,
South, 36, 240; Lancashire,
South-west, 258; Leomins-
ter, 33, 227; Lichfield,
234, 237; Linlithgow, 224;
Londonderry, 259; Co. Londonderry, 223-4; Monmouthshire, 229; Northampton-
shire, 229; Northumberland,
North, 239; Nottingham-
shire, North, 223; Pem-
brokeshire, 28; Perthshire,
245; Preston, 240; Rich-
mond, 230, 236; Ripon,
242, 244; Rye, 224; St
Ives, 227; Sandwich, 223,
244; Scarborough, 230;
Shoreham, 222; Stafford,
232; Staffordshire, North,

STUDIES IN SOCIAL HISTORY

Editor: *HAROLD PERKIN*
Professor of Social History, University of Lancaster

Assistant Editor: *ERIC J. EVANS*
Lecturer in History, University of Lancaster

◇◇◇